FORTSCHRITTE DER PHYSIK

HERAUSGEGEBEN IM AUFTRAGE DER PHYSIKALISCHEN GESELLSCHAFT
DER DEUTSCHEN DEMOKRATISCHEN REPUBLIK
VON F. KASCHLUHN, A. LÖSCHE, R. RITSCHL UND R. ROMPE

27. BAND 1979

AKADEMIE · VERLAG · BERLIN

Inhalt des 27. Bandes

Heft 1

Heft 2/3

Heft 4

Heft 5

Heft 6

Heft 7

Heft 8

Heft 9

Heft 10

Heft 11/12

FORTSCHRITTE DER PHYSIK

HERAUSGEGEBEN IM AUFTRAGE DER PHYSIKALISCHEN GESELLSCHAFT

DER

DEUTSCHEN DEMOKRATISCHEN REPUBLIK

VON F. KASCHLUHN, A. LÖSCHE, R. RITSCHL UND R. ROMPE

HEFT 1 · 1979 · BAND 27

THEORIA CUM PRAXI

AKADEMIE - VERLAG · BERLIN

EVP 10,— M

31728

BEZUGSMÖGLICHKEITEN

Bestellungen sind zu richten

- **in der DDR** an das Zeitungsvertriebsamt, an eine Buchhandlung oder
  an den AKADEMIE-VERLAG,
  DDR - 108 Berlin, Leipziger Str. 3–4
- **im sozialistischen Ausland** an eine Buchhandlung für fremdsprachige Literatur oder an den zuständigen Postzeitungsvertrieb
- **in der BRD und Westberlin** an eine Buchhandlung oder an die Auslieferungsstelle
  KUNST UND WISSEN, Erich Bieber,
  7 Stuttgart 1, Wilhelmstraße 4–6
- **in Österreich** an den Globus-Buchvertrieb,
  1201 Wien, Höchstädtplatz 3
- **in den übrigen westeuropäischen Ländern** an eine Buchhandlung oder an die Auslieferungsstelle
  KUNST UND WISSEN, Erich Bieber GmbH,
  CH - 8008 Zürich/Schweiz, Dufourstraße 51
- **im übrigen Ausland** an den Internationalen Buch- und Zeitschriftenhandel; den Buchexport, Volkseigener Außenhandelsbetrieb der Deutschen Demokratischen Republik,
  DDR - 701 Leipzig, Postfach 160, oder

  an den AKADEMIE-VERLAG,
  DDR - 108 Berlin, Leipziger Str. 3–4

Zeitschrift „Fortschritte der Physik"

Herausgeber: Prof. Dr. Frank Kaschluhn, Prof. Dr. Artur Lösche, Prof. Dr. Rudolf Ritschl, Prof. Dr. Robert Rompe, im Auftrag der Physikalischen Gesellschaft der Deutschen Demokratischen Republik.
Verlag: Akademie-Verlag, DDR - 108 Berlin, Leipziger Straße 3–4; Fernruf: 22 36 221 und 22 36 229; Telex-Nr. 114420; Bank: Staatsbank der DDR, Berlin, Konto-Nr. 6836-26-20712.
Chefredakteur: Dr. Lutz Rothkirch.
Anschrift der Redaktion: Sektion Physik der Humboldt-Universität zu Berlin, DDR - 104 Berlin, Hessische Straße 2.
Veröffentlicht unter der Lizenznummer 1324 des Presseamtes beim Vorsitzenden des Ministerrates der Deutschen Demokratischen Republik.
Gesamtherstellung: VEB Druckhaus „Maxim Gorki", DDR - 74 Altenburg, Carl-von-Ossietzky-Straße 30/31.
Erscheinungsweise: Die Zeitschrift „Fortschritte der Physik" erscheint monatlich. Die 12 Hefte eines Jahres bilden einen Band.
Bezugspreis je Band: 180,– M zuzüglich Versandspesen (Preis für die DDR: 120,– M). Preis je Heft 15,– M (Preis für die DDR: 10,– M).
Bestellnummer dieses Heftes: 1027/27/1.

AN (EDV) 57 618

Fortschritte der Physik 27, 1—34 (1979)

Signal Parameters in Optical Pumping Experiments

L. DELLIT

Department of Physics, Marburg University, Marburg, Federal Republik of Germany

The features of fluorescence and transmission signals in optical pumping of dilute gases with non-coherent light sources are discussed in terms of a generalized signal function. Using the spherical tensor operator formalism kinetic equations for the atomic state multipoles as well as rate equations for spontaneous emission and absorption are derived. In these rate equations the kinetics of the atomic system expressed by the, generally, time dependent state multipoles of the atomic distribution in the ground or excited state is weighted by the optical signal parameters to give a signal function, which is explicitly adapted to the experimental conditions of observation.

Contents

1. Introduction

Commonly, signal detection in optical pumping of dilute gases using noncoherent light sources takes place in two different ways.
One method consists in observing the fluorescence radiation emitted in a certain direction with a certain state of polarization. This radiation, spontaneously emitted by the

atomic system, carries information about the orientation of atoms in the excited state. Since this orientation and the excitation probability depend upon the distribution of atoms in the ground state as well as on the polarisation of the pumping light beam, the fluorescence radiation signal can be used to get informations about the ground state orientation, too.

The other method consists in detecting the transmission of the atomic vapour in a certain polarization channel either for the main pumping light beam or for a separate sample beam. This signal is porportional to the absorption rate and hence carries information about the ground state distribution of the atoms [*1*, *2*].

In both cases and moreover in the description of the optical pumping process itself polarization is one of the most important conceptions. We meet it under different points of view, some of which might the quoted here.

First of all there is the polarization of the pumping light beam that plays a main part in the description of the pumping cycle itself. Under the same scope, the polarization of the signal radiation determines the information about the kinetics of the atomic system in optical detection of the pumping process. Further, together with the investigation of magnetic resonances in optically oriented systems, we use the notion of polarization in order to describe the configuration of magnetic rf-fields. Finally, polarization terms are used in order to characterize the distribution of atoms over their different magnetic substates.

The common scope of all these topics meets with the symmetries of physical parameters under spatial transformations. Therefore, it would be convenient to give the theory in terms of variables with standardized symmetry properties under spatial transformations especially under rotations. This leads to the introduction of spherical tensor operator techniques.

In following this concept it turns out that the geometrical aspect of signal detection in optical pumping experiments can be treated by means of a generalized signal function, which combines the spherical tensor operator components of the polarization operators of the optical detection channel with the tensor operator components of the statistical operator of the atomic system. In this way the kinetics of the atomic system, expressed by the, generally, time dependent state multipoles of the atomic distribution in the ground or excited state, is weighted by the optical signal parameters to give a signal function, which is explicitly adapted to the experimental conditions of observation.

This, however, is only one aspect under which the introduction of the spherical tensor operator formalism into the description of the optical pumping process seems to be convenient.

The other ones account to the fact, that using the apriori symmetries of the interaction between atoms and the electromagnetic field or else the symmetries of relaxation interactions, one is led to a useful separation of the basic dynamical parameters from spatial symmetry problems. The kinetic equations for the state multipoles of the atomic system forming, in general, an interdependent system of linear differential equations with time dependent coefficients, then probably decouple in a fashion, which is clearly to survey once the basic symmetries of the interaction in question is determined [*3*].

Actually, there is a strong interdependence between the description of the optical pumping process itself and the description of its optical detection methods, since both can be regarded as different aspects in solving the equation of motion for the global statistical operator of the combined system of atoms and the radiation field. Using the "regressive method" informations about the matter system or the radiation field can be obtained by taking the partial trace of the global density matrix over the radiation variables or matter variables, respectively. A brief survey on the main features of this conception was given in earlier papers [*4*, *5*].

Meanwhile, in a quite similar fashion, a rigourous and detailed treatment in terms of

kinetic equations starting with the quantum-mechanical Liouville equation for the global density operator was published by WILLIS [6]. The linearized form of this theory leads to essentially the same results as reported in reference [4] and [5]. The kinetic equation for the single particle density operator of the matter system gives the equations of COHEN-TANNOUDJI [2] in an extended notation using a generalized polarization matrix. Following this concept, our handling with the problem splits into two parts: the introduction of spherical tensor notation into the description of the optical pumping process and, based on this same formalism, an outline of the features of optical detection signals.
In a first part (Sec. 2 through 3.5) we derive kinetic equations for the state multipoles of the atomic system. Here we could start with the well known equations of COHEN-TANNOUDJI, but there are the following three aspects which suggest a more detailed discussion.
First, in Sec. 2, we sketch the common starting point for the description of optical pumping and its optical detection formulated by the Liouville eqution for the global statistical operator. In their linearized form the approximate solutions of the Liouville equation meet with the two scopes of optical pumping and of its optical detextion. Secondly, in Sec. 3.1, we introduce the statistical operator technique in handling with the polarization of the radiation field which, not at least, turns out to be useful with respect to the spherical tensor notation. Finally, in Sec. 3.2, we have to discuss some essentials which occur in the derivation of COHEN-TANNOUDJI's equations, such as the electric dipole approximation and the broadband condition, in order to point out their relation to the advantages of the spherical tensor formalism.
In a second part (Sec. 4 through 5.4) we then derive the basic equations for optical signal detection. Using the same methods set forth above, we introduce the concept of a generalized signal function and discuss some features with respect to the various conditions of observation. This can be seen as an extension of former results quoted by D'YAKONOV [7] concerning the multipole character of fluorescence radiation in optical pumping experiments.
In order to avoid confusion concerning phase conventions the main properties of the spherical tensor operators used in this context are summarized in the appendix.

2. The Liouville equation for the atom-optical field system. Approximate solution

The statistical operator — density operator or density matrix — $\mu(t)$ for a system of N atoms interacting with an electromagnetic radiation field satisfies the Liouville equation:

$$i\hbar\, d\mu(t)/dt = [\mathscr{H}, \mu(t)], \tag{1}$$

where $\mathscr{H}$ is the Hamiltonian of the radiation-matter system:

$$\begin{aligned}\mathscr{H} &= H_N + H_f + \eta V_N \\ &= H_0(N, \{\boldsymbol{k}\}) + \eta V_N\end{aligned} \tag{2}$$

and

$$H_N = \sum_{n=1}^{N} {}^{(n)}H$$

$$H_f = \hbar c \sum_{\boldsymbol{k},\sigma} |\boldsymbol{k}|\, a^{+}_{\boldsymbol{k},\sigma} a_{\boldsymbol{k},\sigma}$$

$$V_N = \sum_{n=1}^{N} {}^{(n)}V = \sum_{n=1,\boldsymbol{k},\sigma} (a^{(n)}_{\boldsymbol{k},\sigma} V_{\boldsymbol{k},\sigma} + \text{h.c.}).$$

$^{(n)}H$ is the Hamiltonian of the n-th atom[1]). The interaction $^{(n)}V$ of the n-th atom with the radiation field couples the atomic ground state $|a_{(n)}, \alpha_{(n)}\rangle$ and the excited state $|b_{(n)}, \beta_{(n)}\rangle$ where η (cf. (27)) is the coupling constant. For a single atom we have:

$$H\,|a, \alpha\rangle = E_\alpha\,|a, \alpha\rangle$$
$$H\,|b, \beta\rangle = E_\beta\,|b, \beta\rangle. \tag{3}$$

In this notation α, β stands for the magnetic quantum number of the ground and excited state respectively whereas a, b denotes a set of further quantum numbers particularly the quantum number j of total angular momentum. If it is not convenient to specify the ground and excited state we denote two different atomic eigenstates by $|j, m\rangle$ and $|j', m'\rangle$.

As a complete set of orthonormal basis functions for the electromagnetic radiation field we take plane waves normalized within a cavity with periodic boundary conditions. Each mode $M_{\boldsymbol{k},\sigma}$ of the cavity is characterized by a wave vector $\boldsymbol{k}$ and an index σ corresponding to any two linearly independent states of polarization. The $a_{\boldsymbol{k},\sigma}$ and $a^+_{\boldsymbol{k},\sigma}$ are the annihilation and creation operators of a photon in the mode $M_{\boldsymbol{k},\sigma}$.

The statistical operator μ is defined over a state space $\mathscr{E}$ being the tensor product of the spaces $^{(n)}\mathscr{E}$ and $\mathscr{E}_{\boldsymbol{k},\sigma}$ of all the N atoms and the enumerable but infinite number of radiation modes.

The Hamiltonian H_N of Eq. (2) is the Hamiltonian of N non-interacting atoms. If we specify this model to the conditions of optical pumping we assume the statistical operator μ to be a symmetric function in the matter variables. That means we consider a system of N atoms of the same kind each of them interacting with the same radiation field represented by the statistical operator μ_{rad}. We are then interested in the dynamics of a single atom taking the average over the N-particle system which leads to the single particle statistical operator μ_{at}. One defines:

$$\mu_{\text{rad}} = \text{tr}_{1,2,\ldots,N}\,\mu$$
$$^{(1)}\mu_{\text{at}} = \text{tr}_{2,3,\ldots,N}\,\text{tr}_{\text{rad}}\,\mu \tag{4}$$

where $\text{tr}_{1,2,\ldots,N}$ indicates the trace over the variables of the first, second, ..., N-th atom and tr_{rad} stands for the trace over all the radiation modes. Assuming μ to be symmetric in the matter variables the special manner of enumerating the different atoms is of no account. Hence:

$$^{(1)}\mu_{\text{at}} = {}^{(2)}\mu_{\text{at}} = \cdots \equiv \mu_{\text{at}}.$$

With the definitions (4) and together with:

$$\hat{\mu}(t) = \text{tr}_{1,2,\ldots,n-1,n+1,\ldots,N}\,\mu(t) \tag{5}$$

one deduces from (1) kinetic equations for μ_{rad} and μ_{at}:

$$i\hbar\dot{\mu}_{\text{rad}}(t) = [H_f, \mu_{\text{rad}}(t)] + N\eta\,\text{tr}_{\text{at}}\,[V, \hat{\mu}(t)]$$
$$i\hbar\dot{\mu}_{\text{at}}(t) = [H, \mu_{\text{at}}(t)] + \eta\,\text{tr}_{\text{rad}}\,[V, \hat{\mu}(t)] \tag{6}$$

[1]) For simplicity we neglect the energy of the classical center-of-mass motion of the atoms.

where H and V without an index indicate the single particle Hamiltonian and interaction operator ($H = {}^{(1)}H$, $V = {}^{(1)}V$). The kinetic equations are coupled by $\hat{\mu}(t)$ which describes the single atom-radiation correlations[2]).
An approximate solution of the Liouville equation in the form (6) starts with an approximate solution for $\hat{\mu}(t)$ assuming that the interaction energy ηV can be treated as a small perturbation and that the matter and the radiation systems are uncorrelated at $t = 0$. The evolution in time of $\hat{\mu}(t)$ is given by:

$$\hat{\mu}(t) = U(t, 0)\, \hat{\mu}(0)\, U^{+}(t, 0)$$

where $U(t, 0)$ is the solution of the differential equation:

$$i\hbar \dot{U}(t, 0) = \big(H_0(n, \{\boldsymbol{k}\}) + \eta V\big)\, U(t, 0).$$

Using the equations for the unperturbed system:

$$\hat{\mu}_0(t) = U_0(t, 0)\, \hat{\mu}(0)\, U_0^{+}(t, 0)$$

and

$$i\hbar \dot{U}_0(t, 0) = H_0(n, \{\boldsymbol{k}\})\, U_0(t, 0)$$

we can write $\hat{\mu}(t)$ in the form:

$$\hat{\mu}(t) = W(t, 0)\, \hat{\mu}_0(t)\, W^{+}(t, 0) \tag{7}$$

with

$$W(t, 0) \doteq U(t, 0)\, U_0^{+}(t, 0).$$

Integrating the resulting differential equation for the operator $U_0^{+} W U_0$ one obtains for $W(t, 0)$:

$$W(t, 0) = 1 - (i/\hbar)\, \eta \mathscr{L}(t) \tag{8}$$

with

$$\mathscr{L}(t) = \int_0^t U_0(t, \tau)\, V U_0^{+}(t, \tau)\, d\tau.$$

Inserting (8) into (7) we obtain to the first order in η:

$$\hat{\mu}(t) = \hat{\mu}_0(t) - (i/\hbar)\, \eta [\mathscr{L}(t), \hat{\mu}_0(t)] \tag{9}$$

and with the assumption that the matter and the radiation systems are uncorrelated at $t = 0$ we have:

$$\hat{\mu}_0(t) = \mu_{\text{at}}(t)\, \mu_{\text{rad}}(t). \tag{10}$$

Using these results together with (6) we can write:

$$i\hbar\, \dot{\mu}_{\text{rad}}(t) = [H_f, \mu_{\text{rad}}(t)] - (i/\hbar)\, \eta^2\, N \operatorname{tr}_{\text{at}} \big[V, [\mathscr{L}(t), \mu_{\text{at}}(t)\, \mu_{\text{rad}}(t)]\big] \tag{11a}$$

$$i\hbar \dot{\mu}_{\text{at}}(t) = [H, \mu_{\text{at}}(t)] - (i/\hbar)\, \eta^2 \operatorname{tr}_{\text{rad}} \big[V, [\mathscr{L}(t), \mu_{\text{at}}(t)\, \mu_{\text{rad}}(t)]\big] \tag{11b}$$

where terms linear in the interaction V and hence linear in the $a_{\boldsymbol{k},\sigma}$ are omitted. They give no contribution to the dynamics under the conditions of a light beam with thermal-like statistics since their expectation values vanish (see f.i. [*6*, *11*]).
Eqs. (11) constitute the fundamental set of kinetic equations for optical pumping and its optical detection reported in a slightly different notation in ref. [*4*, *5*]. The reader who is interested in a detailed discussion of some further assumption sconcerning the limits of integration in (8) may be referred to Ref. [*2*] and [*6*]. The so called "repeated random-

[2]) $\hat{\mu}(t)$ likewise depends by a similar set of equations on the higher order correlations and finally depends on $\mu(t)$ [*6*].

phase" assumption states, that the time $t = 0$ in the derivation of (9) is optional and an investigation of relaxation and interaction time constants shows that the upper limit of integration in (8) could be extended to infinity.
Presumably the easiest approach to a linerarized form of the theory is to restrict the consideration to a situation where one can neglect the reaction of the matter system on the pumping light beam. Hence we solve (11b) for the single particle statistical operator μ_{at} assuming μ_{rad} to be essentially the statistical operator of the incident pumping light and to be constant over the pumping region. These exerimental conditions are widely fulfilled in optical pumping of rarefied gases. Different situations were described i.e. by BARRAT [8] and COHEN-TANNOUDJI and LALOE [9] including multiple scattering and variations of the pumping light beam propagating in an atomic vapour.
On the other hand, when solving (11a) in order to get the rate equations for spontaneous emission and absorption we consider μ_{rad} as to be the statistical operator of the fluorescence light or to be the statistical operator of the sample beam in a transmission experiment, assuming μ_{at} to be a given solution of (11b).
Finally, the thermal character of the radiation field is a striking point in our further considerations. In the following chapter we give a brief outline of the main properties of μ_{rad} which hold for a radiation field with thermal-like statistics.

3. Kinetic equations for the atomic state multipoles

3.1. The statistical operator of the pumping light beam

As mentioned in the foregoing section we take plane waves to form a complete set of orthonormal basis functions for the radiation field normalized within a volume V_0 with periodic boundary conditions. Thus for instance the operator for the radiation-gauge vector potential $\boldsymbol{A}$ is given by the expansion

$$\boldsymbol{A}(\boldsymbol{r}) = \sum_{\boldsymbol{k},\sigma} (\hbar/2\varepsilon_0\omega_k)^{1/2} \left(a_{\boldsymbol{k},\sigma}\mathfrak{a}_{\boldsymbol{k},\sigma}(\boldsymbol{r}) + a^{+}_{\boldsymbol{k},\sigma}\mathfrak{a}^{*}_{\boldsymbol{k},\sigma}(\boldsymbol{r})\right) \tag{12}$$

with

$$\mathfrak{a}_{\boldsymbol{k},\sigma}(\boldsymbol{r}) = (V_0)^{-1/2}\, \boldsymbol{e}_\sigma(\boldsymbol{k}) \exp\,(i\boldsymbol{k}\cdot\boldsymbol{r})$$

$$\omega_k = k(\varepsilon_0\mu_0)^{-1/2} = k\cdot c\,.$$

To describe the polarization labeled by the quantum number σ we choose a spherical basis representing circularly polarized photons. Let $\boldsymbol{e}_1(\boldsymbol{k})$ and $\boldsymbol{e}_2(\boldsymbol{k})$ be any two real unit vectors forming together with the wave vector $\boldsymbol{k}$ a right-handed Cartesian triad. One then defines [*10*]:

$$\begin{aligned} \boldsymbol{e}_\sigma(\boldsymbol{k}) &= -\frac{\sigma}{\sqrt{2}}\left(\boldsymbol{e}_1(\boldsymbol{k}) + i\sigma\boldsymbol{e}_2(\boldsymbol{k})\right) \\ \boldsymbol{e}_0(\boldsymbol{k}) &= \boldsymbol{k}/k = \boldsymbol{e}_3(\boldsymbol{k}), \qquad \sigma = +1, -1. \end{aligned} \tag{13}$$

Tor a photon in the polarization mode $\boldsymbol{e}_\sigma(\boldsymbol{k})$ the component of angular momentum along Fs direction of propagation is $\sigma\hbar$.
ito specify our notion of an optical radiation field with thermal-like statistics we assume that each mode $M_{\boldsymbol{k},\sigma}$ of the field is coupled to a number of sources via a stochastic process which ist both, stationalry and ergodic. Such sources might, in practice, be several statistically independent subdivision of one large source. In the Fock respresentation,

the statistical operator (of the Gaussian type[3]) for the mode $M_{\boldsymbol{k},\sigma}$ is:

$$\mu_{\boldsymbol{k},\sigma} = \sum_{m=0}^{\infty} \langle n_{\boldsymbol{k},\sigma}\rangle^m \, (1 + \langle n_{\boldsymbol{k},\sigma}\rangle)^{-(m+1)} \, |m, \boldsymbol{k}, \sigma\rangle \, \langle m, \boldsymbol{k}, \sigma| \tag{14}$$

where

$$\langle n_{\boldsymbol{k},\sigma}\rangle = \mathrm{tr}\,(\mu_{\boldsymbol{k},\sigma} a^{+}_{\boldsymbol{k},\sigma} a_{\boldsymbol{k},\sigma})$$

is the average photon number in the mode $M_{\boldsymbol{k},\sigma}$ and

$$|m, \boldsymbol{k}, \sigma\rangle = (a^{+}_{\boldsymbol{k},\sigma})^m \, (m!)^{-1/2} \, |\mathrm{vac}\rangle \tag{15}$$

is the m-quantum state of this mode [*11*, *12*].

The states $|m, \boldsymbol{k}, \sigma\rangle$ form a complete orthonormal set spanning the state space $\mathscr{E}_{\boldsymbol{k},\sigma}$ and in the Fock space being the tensor product of the infinite but enumerable spaces $\mathscr{E}_{\boldsymbol{k},\sigma}$ we have:

$$\mu_{\mathrm{rad}} = \prod_{\boldsymbol{k},\sigma} \otimes \, \mu_{\boldsymbol{k},\sigma} \, . \tag{16}$$

The stationarity of the field represented by (16) is expressed by the fact that μ_{rad} commutes with the Hamiltonian H_f of the field (cf. (2)). Thus the statistical statements we make concerning the pumping light beam are invariant under time displacements. In this model the average photon numbers $\langle n_{\boldsymbol{k},\sigma}\rangle$ for instance are assumed to be time-independent.

In the spirit of the usual passage to the limit $V_0 \to \infty$ the distribution of states in wave vector space is continuous and the expectation values $\langle n_{\boldsymbol{k},\sigma}\rangle$ are related to the spectral energy flux density $c \cdot dE(\boldsymbol{k}, \sigma)$ of radiation within the frequency intercal $(\omega, \omega + d\omega)$, the solid angle $(\Omega, \Omega + d\Omega)$ and with polarization σ according to:

$$c \cdot dE(\boldsymbol{k}, \sigma) = I(\boldsymbol{k}, \sigma) \, d\omega \, d\Omega = (2\pi)^{-3} \, \hbar c k^3 \langle n_{\boldsymbol{k},\sigma}\rangle \, d\omega \, d\Omega \, . \tag{17}$$

For what follows it is an important fact that the average number of photons in a single mode obtained with thermal light usually used in optical pumping experiments is sufficiently low [*13*] so that expansions such as (14) need only be extended up to the first order in the average photon number. For the fully circularly polarized field with wave vektor $\boldsymbol{k}$ we have according to (14):

$$\mu_{\boldsymbol{k},\sigma} = (1 - \langle n_{\boldsymbol{k},\sigma}\rangle) \, |0, \boldsymbol{k}\rangle \, \langle 0, \boldsymbol{k}| + \langle n_{\boldsymbol{k},\sigma}\rangle \, \pi_{\boldsymbol{k},\sigma} \tag{18}$$

where

$$\pi_{\boldsymbol{k},\sigma} = |1, \boldsymbol{k}, \sigma\rangle \, \langle 1, \boldsymbol{k}, \sigma|$$

is the projector onto the one-quantum state of the mode $M_{\boldsymbol{k},\sigma}$.

Let us briefly consider two examples in order to illustrate the handling with statistical mixtures.

1) Usually the random type of excitation leads to an independent but equal occupation of the two polarization modes and the radiation delivered by the lamp is unpolarized. Then according to (16) we have to consider the direct products $\mu_{\boldsymbol{k},\sigma} \otimes \mu_{\boldsymbol{k},\sigma'}$. Expanding these products up to the first order in the average photon number we have for the wave vector $\boldsymbol{k}$:

$$\mu_{\boldsymbol{k},\mathrm{unpol}} = (1 - \langle n_{\boldsymbol{k}}\rangle) \, |0, \boldsymbol{k}\rangle \, \langle 0, \boldsymbol{k}| + \langle n_{\boldsymbol{k}}\rangle \, \pi_{\boldsymbol{k},\mathrm{unpol}} \tag{19}$$

with

$$\pi_{\boldsymbol{k},\mathrm{unpol}} = (1/2) \, (|1, \boldsymbol{k}, \sigma\rangle \, \langle 1, \boldsymbol{k}, \sigma| + |1, \boldsymbol{k}, \sigma'\rangle \, \langle 1, \boldsymbol{k}, \sigma'|)$$

---

[3]) The random sort of excitation presumably characteristic of most of the familiar types of non-coherent macroscopic light sources, such as gas discharges, leads to a Gaussian distribution for the weight function in a coherent state representation.

and

$$\langle n_{\boldsymbol{k}} \rangle = \langle n_{\boldsymbol{k},\sigma} \rangle + \langle n_{\boldsymbol{k},\sigma'} \rangle = 2\langle n_{\boldsymbol{k},\sigma} \rangle .$$

2) If we are intersted in experiments with a linearly polarized beam we can describe the effect of a polarizer by subjecting the incoherent mixture (19) to a certain preparation measurement. According to (13), let $\boldsymbol{e}_\varphi(\boldsymbol{k})$ be an unit vector perpendicular to $\boldsymbol{k}$:

$$\boldsymbol{e}_\varphi(\boldsymbol{k}) = \boldsymbol{e}_1(\boldsymbol{k}) \cos \varphi + \boldsymbol{e}_2(\boldsymbol{k}) \sin \varphi . \tag{20}$$

Then one-quantum state of this mode is:

$$|1, \boldsymbol{k}, \varphi\rangle = |1, \boldsymbol{k}, \sigma\rangle \big(\boldsymbol{e}_\sigma(\boldsymbol{k})\, \boldsymbol{e}_\varphi(\boldsymbol{k})\big) + |1, k, \sigma'\rangle \big(\boldsymbol{e}_{\sigma'}(\boldsymbol{k})\, \boldsymbol{e}_\varphi(\boldsymbol{k})\big) \tag{21}$$

and the preparation measurement is represented by the operator:

$$\mathscr{P}_{\boldsymbol{k},\sigma} = |0, \boldsymbol{k}\rangle \langle 0, \boldsymbol{k}| + |1, \boldsymbol{k}, \varphi\rangle \langle 1, \boldsymbol{k}, \varphi| .$$

After passing the polarizer which is assumed to have ideal polarization characteristics the statistical operator of the field with wave vector $\boldsymbol{k}$ is given by (cf. [*10*]):

$$\mu_{\boldsymbol{k},\varphi} = (\mathscr{P}_{\boldsymbol{k},\varphi}\, \mu_{\boldsymbol{k}} \mathscr{P}_{\boldsymbol{k},\varphi}) / \mathrm{tr}\, (\mu_{\boldsymbol{k}} \mathscr{P}_{\boldsymbol{k},\varphi}) = (1 - \langle n_{\boldsymbol{k},\varphi} \rangle)\, |0, \boldsymbol{k}\rangle \langle 0, \boldsymbol{k}| + \langle n_{\boldsymbol{k},\varphi} \rangle\, \pi_{\boldsymbol{k},\varphi} \tag{22}$$

with

$$\pi_{\boldsymbol{k},\varphi} = |1, \boldsymbol{k}, \varphi\rangle \langle 1, \boldsymbol{k}, \varphi|$$

and in the special case of an input according to (19):

$$\langle n_{\boldsymbol{k},\varphi} \rangle = \langle n_{\boldsymbol{k}} \rangle / 2 .$$

It is evident that essentially the same result would be obtained if we started with a basis of linearly polarized mode functions regarding (13) as an unitary basis transformation. Let us conclude this section with a remark concerning a reasonable beam model.
The mechanism of optical pumping is often described as a tranfser of angular momentum, or more generally, as a transfer of spatial symmetry from the optical radiation field to the atomic system. Hence a preferable arrangement would contain a pumping light beam of photons with definite angular momentum or spatial symmetry. In the spirit of a simplified model we therefore can assume the components of the beam differing only according to the spectral function $\langle n(\omega) \rangle$, having a unified direction of propagation along the unit vector $\boldsymbol{e}$ and the same polarization matrix $\langle \sigma | \pi_{\boldsymbol{e}} | \sigma' \rangle$. The statistical properties of the pumping light beam would therefore be uniquely determined by the data set

$$\boldsymbol{e},\ \langle n(\omega) \rangle,\ \langle \sigma | \pi_{\boldsymbol{e}} | \sigma' \rangle \tag{23}$$

where $\langle \sigma | \pi_{\boldsymbol{e}} | \sigma' \rangle$, the matrix element of the statistical operator $\pi_{\boldsymbol{k}}$ with $\boldsymbol{k} = (\omega/c)\, \boldsymbol{e}$, is independent of ω.

3.2. Kinetic equations for the atomic density matrix

In order to calculate the evolution in time of μ_{at} caused by interaction with the optical radiation field according to (11b) one has to specify further assumptions concerning the interaction operators V and $\mathscr{L}$. For a particular matrix element of $\mathscr{L}$ according to (8) we find for instance:

$$\langle a, \alpha; 1, \boldsymbol{k}, \sigma | \mathscr{L} | b, \beta \rangle = \langle 1, \boldsymbol{k}, \sigma | a^+_{\boldsymbol{k},\sigma} | 0, \boldsymbol{k} \rangle \langle a, \alpha | V^+_{\boldsymbol{k},\sigma} | b, \beta \rangle \cdot 2\pi \delta_-\big(\omega_k - \omega_0(\beta, \alpha)\big) \tag{24}$$

where

$$\delta_-(x) = 1/(2\pi) \int_0^\infty \exp\,(-ixt)\, dt = (1/2)\, (\delta(x) - (i/\pi)\ \mathrm{PP}\!\int (1/x)\, dx)$$

and

$$\omega_0(\beta, \alpha) = \hbar^{-1}(E_{b,\beta} - E_{a,\alpha}).$$

Writing down these equations we made the following two assumptions:

a) We have assumed that the interaction with the radiation field induces transitions where one photon is absorbed or emitted conserving the energy of the unperturbed system within the natural line width. The same arguments which led to the utmost one quantum state assumption stated above hold for neglecting induced emission processes.

b) Furthermore, absorption and spontaneous emission can be treated perturbatively. As for the absorption process this means that the line width of the pumping radiation is much greater than the inverse lifetime Γ_p of the atomic ground state (broad band condition) and for the spontaneous emission it means that the energetic distance ω_0 in the frequency scale between the ground and excited state of the atom is much greater than the inverse lifetime Γ_s of the excited state (WIGNER-WEISSKOPF-approximation).

As a consequence of the thermal-like character of the field it can be shown that the stationary solutions of (11b) lead to vanishing matrix elements of μ_{at} between the ground and excited states. This is a main feature leading to a useful decomposition in the statistical operator μ_{at} (cf. (38)). Since the corresponding deduction is easy to survey we give a brief outline of this calculation in order to point out some general aspects which are important when introducing spherical tensor notation.

Calculating the time-derivative of the matrix elements between the ground and the excited states $\langle a, \alpha| \mu_{\text{at}} |b, \beta\rangle$ according to (11b) there are only the products $\mu_{\text{at}}\mu_{\text{rad}}\mathscr{L}V$ and $V\mathscr{L}\mu_{\text{at}}\mu_{\text{rad}}$ which give a non-vanishing contribution related to spontaneous emission and absorption, respectively. Taking the trace over the field variables and using (24) we obtain for these products:

$$\begin{aligned}&\langle a, \alpha| \operatorname{tr}_{\text{rad}}(\mu_{\text{at}}\mu_{\text{rad}}\mathscr{L}V) |b, \beta\rangle \\ &= 2\pi \sum_{\substack{\boldsymbol{k},\sigma \\ \alpha',\beta'}} (1 - \langle n_{\boldsymbol{k}}\rangle)\, \delta_-\big(\omega_0(\beta', \alpha') - \omega_k\big) \langle b, \beta'| V_{\boldsymbol{k},\sigma} |a, \alpha'\rangle \langle a, \alpha'| V^+_{\boldsymbol{k},\sigma} |b, \beta\rangle \langle a, \alpha| \mu_{\text{at}} |b, \beta'\rangle\end{aligned}$$

emission) (25a)

$$\begin{aligned}&\langle a, \alpha| \operatorname{tr}_{\text{rad}} (V\mathscr{L}\mu_{\text{at}}\mu_{\text{rad}}) |b, \beta\rangle \\ &= 2\pi \sum_{\substack{\boldsymbol{k},\sigma,\sigma' \\ \alpha',\beta'}} \langle n_{\boldsymbol{k}}\rangle \langle\sigma'| \pi_{\boldsymbol{k}} |\sigma\rangle\, \delta_-\big(\omega_0(\beta',\alpha') - \omega_k\big) \langle a,\alpha| V^+_{\boldsymbol{k},\sigma} |b,\beta'\rangle \langle b,\beta'| |V_{\boldsymbol{k},\sigma'} |a,\alpha'\rangle \langle a,\alpha'| \mu_{\text{at}} |b,\beta\rangle\end{aligned}$$

(absorption). (25b)

Spontaneous emission:

In the emission term (25'a) we neglect $\langle n_{\boldsymbol{k}}\rangle$ compared to 1 and in the limit $V_0 \to \infty$, the spacing in $\boldsymbol{k}$ becomes arbitrarily small. The sum over $\boldsymbol{k}$ changes into an integral according to the scheme:

$$\sum_{\boldsymbol{k}} \to V_0(2\pi)^{-3} \int k^2\, dk\, d\Omega. \tag{26}$$

Integrating over the solid angle Ω we have to pay attention to the angular dependence of the interaction matrix elements. Since here we meet an essential of further argumentation we should consider the interaction operator in some more detail.

In the non-relativistic approximation, the electron-radiation coupling Hamiltonian up to the first order in the vector potential is:

$$H_{\text{int}} = \eta V = -(e/m) \sum_i \big(\boldsymbol{p}_i\boldsymbol{A}(\boldsymbol{r}_i) + g_i(\hbar/2)\, \boldsymbol{s}_i \operatorname{curl} \boldsymbol{A}(\boldsymbol{r}_i)\big),$$

$\boldsymbol{P}_i$ being the momentum, $\boldsymbol{s}_i$ the spin and g_i the g-factor of the i-th electron. Inserting for the vector potential according to (12) and using the Rayleigh expansion of plane waves into multipole fields [*14*] we can write $V_{\boldsymbol{k},\sigma}$ in the form:

$$V_{\boldsymbol{k}\,\sigma} = \sum_{K,i} \left(\mathscr{E}^{(K)}_{\sigma,i}(\boldsymbol{k}) + \mathscr{M}^{(K)}_{\sigma,i}(\boldsymbol{k})\right),$$

where the multipole operators $\mathscr{E}^{(K)}_{\sigma,i}(\boldsymbol{k})$ or $\mathscr{M}^{(K)}_{\sigma,i}(\boldsymbol{k})$, respectively, represent an electric or magnetic moment of order K, related to the direction $\boldsymbol{k}$. They are components of irreducible tensor operators of order K, which transform under rotation according to (A 1). The contribution of the different terms in the multipole expansion of $V_{\boldsymbol{k},\sigma}$ is mainly determined by the selection rules for angular momentum and parity. In the case of optical transitions the long wavelength approximation is adopted and electric dipole translations account for the overwhelming majority of spectral lines. Under these conditions, explicit expressions for the coupling constant and the matrix elements of $V_{\boldsymbol{k},\sigma}$ are:

$$\eta = \left(\hbar/(2\varepsilon_0 V_0)\right)^{1/2} \tag{27}$$

and

$$\langle j, m|\; V_{\boldsymbol{k},\sigma}\; |j', m'\rangle = -i\omega_{j,m;j',m'}(\omega_k)^{-1/2}\, \langle j, m|\; \boldsymbol{D}\boldsymbol{e}_\sigma(\boldsymbol{k})\; |j', m'\rangle$$

with

$$\omega_{j,m;j',m'} = \hbar^{-1}(E_{j,m} - E_{j',m'}),$$

where $\boldsymbol{D}$ is the operator of the electric dipole moment of the atom.
From the foregoing it is clear that $V_{\boldsymbol{k},\sigma}$, in this case, is proportional to the σ-component of an irreducible tensor operator of order 1. Let us illustrate this statement by the following remarks gathering the main results of this excursus.
If we relate the spherical basis $\boldsymbol{e}_\sigma(\boldsymbol{k})$ to a cartesian triad $\boldsymbol{e}_i$ ($i = 1, 2, 3$) which is rather fixed to the atomic system than to the direction of the pumping beam, we must take into account the transformation:

$$\boldsymbol{e}_\lambda(\boldsymbol{k}) = \sum_{\nu} \boldsymbol{e}_\nu R^{(1)}_{\nu,\lambda}(\boldsymbol{k}) \qquad \lambda, \nu = 0, \pm 1, \tag{28}$$

where the rotation $\mathscr{R}(\boldsymbol{k})$ transforms the triad $\boldsymbol{e}_i$ into the triad $\boldsymbol{e}_i(\boldsymbol{k})$ introduced with the notation (13) (cf. Sec. 5.4.). Hence we have

$$\boldsymbol{D}\boldsymbol{e}_\sigma(\boldsymbol{k}) = \sum_{\lambda} D_\lambda{}^{(1)} R^{(1)}_{\lambda,\sigma}(\boldsymbol{k}) \qquad \sigma = \pm 1; \qquad \lambda = 0, \pm 1 \tag{29}$$

and for the matrix element of the irreducible tensor operator component $D_\lambda{}^{(1)} = \boldsymbol{D}\boldsymbol{e}_\lambda$ we have according to the WIGNER-ECKART theorem:

$$\langle j, m|\; D_\lambda{}^{(1)}\; |j', m'\rangle = (-1)^{j'-m'}\, \hat{1}^{-1} \langle j, m; j', -m' \mid 1, \lambda\rangle\, \langle j \|\, D^{(1)}\, \| j'\rangle \tag{30}$$

where $\langle j, m; j', -m' \mid 1, \lambda\rangle$ is a Clebsch-Gordan coefficient, $\langle j\|\, D^{(1)}\, \|j'\rangle$ the reduced matrix element and where we used the abbreviation $\hat{j} = (2j + 1)^{1/2}$ [*10*][4]).
Let us now return to the matrix elements of μ_{at} between the ground and the excited states. Calculating (25a) with the assumptions stated above and inserting into (11b) we finally obtain for the contribution due to spontaneous emission:

$$\langle a, \alpha|\; \dot{\mu}_{\text{at}}\; |b, \beta\rangle_{\text{sp.em.}} = -\left((2T_s)^{-1} - i\hbar^{-1}\,\Delta E\right) \langle a, \alpha|\; \mu_{\text{at}}\; |b, \beta\rangle \tag{31}$$

[4]) Diverging from [*10*] we use an obvious rearrangement in the notation of Clebsch-Gordan coefficients according to EDMONDS [*15*].

with

$$1/T_s = \Gamma_s = (3\pi\varepsilon_0\hbar)^{-1} \sum_\lambda \int k^2\, dk\, \delta(\omega_0 - \omega_k)\, (\omega_0^2/\omega_k)\, |\langle a, \alpha'|\, D_\lambda^{(1)}\, |b, \beta\rangle|^2 \tag{32}$$

and

$$\Delta E = (6\pi^2\varepsilon_0)^{-1} \sum_\lambda \mathrm{PP}\int k^2\, dk (\omega_0^2/\omega_k)\, (\omega_0 - \omega_k)^{-1}\, |\langle a, \alpha'|\, D_\lambda^{(1)}\, |b, \beta\rangle|^2,$$

where we used the relation:

$$\sum_\sigma \int d\Omega R^{(1)}_{\lambda,\sigma} R^{(1)*}_{\lambda',\sigma} = (8\pi/3)\, \delta_{\lambda,\lambda'}. \tag{33}$$

T_s is the mean lifetime and ΔE the 'level shift' [*10*] of the excited state of the atom. Strictly speaking, the lifetime T_s and the level shift ΔE would, generally, depend upon the quantum number β of the excited sublevel via ω_0 according to (24). So the expressions (32) are correct only in the case of energetic degeneracy as well in the ground as in the excited states. On the other hand, often in optical pumping experiments this degeneracy is removed by interaction with an external magnetic field. But we assume the Zeeman level shifts to be much smaller than the energetic distance between the ground and the excited states of the atom so that dealing with the unified relations (32) is a good approximation. We will return to this point in connection with the absorption process.

Absorption:

In case of the absorption term (25b), the integration over $\boldsymbol{k}$-space would essentially be determined by the structure of the pumping light beam expressed by the function $\langle n_{\boldsymbol{k}}\rangle \langle\sigma'|\, \pi_{\boldsymbol{k}}\, |\sigma\rangle$. Beside the Ω-dependence related to the beam divergence, the spectral function $\langle n_{\boldsymbol{k}}\rangle$ and polarization parameters $\langle\sigma'|\, \pi_{\boldsymbol{k}}\, |\sigma\rangle$, in general, will not be independent and the interaction with the radiation would transfer a great deal of spatial unsymmetries into the atomic system. In our context, however, is should be pointed out that the advantages of introducing irreducible tensor notation were missed in the general case. This is the very reason which justifies the simplified model for the pumping light beam according to (23). On the other hand, this approximation is not too bad and is widely used in describing optical pumping experiments.

Usually, for convenience, one assumes the axis of quantization for the atomic system to coincide with the direction of propagation of the pumping light beam. This convenience might be missed in the general case, where the presence of further interactions, i.e. relaxation processes, probably introduce different axes of symmetry into the problem. We treat the general case inserting the transformation formulae (29) for the beam direction $\boldsymbol{e}$ into (25b) and combine the rotation matrix elements and polarization parameters according to:

$$\langle\lambda'|\, \pi\, |\lambda\rangle = \sum_{\sigma,\sigma'} R^{(1)}_{\lambda',\sigma'}(\boldsymbol{e})\, \langle\sigma'|\, \pi_{\boldsymbol{e}}\, |\sigma\rangle\, R^{(1)*}_{\lambda,\sigma}(\boldsymbol{e})$$
$$\sigma, \sigma' = \pm 1 \qquad \lambda, \lambda' = 0, \pm 1. \tag{34}$$

Thus we define a polarization operator π which is the transform of $\pi_{\boldsymbol{e}}$ under the rotation $\mathscr{R}(\boldsymbol{e})$ introduced together with (28).

This interpretation of the transformation rules for the polarization parameters by means of the transformation rules for the statistical operator $\pi_{\boldsymbol{e}}$ is not trivial. It could be done because the interaction is represented by a first order spherical tensor operator $\boldsymbol{D}^{(1)}$ and $\pi_{\boldsymbol{e}}$ can be treated as an operator acting in a state space of total angular momentum quantum number $J = 1$. This situation is different from those in nuclear transitions where multipole matrix elements of order greater than 1 are to be taken into account. But sinc the electric dipole approximation is a good one for the description of optical transitions the convenience of the above interpretation justifies the reduction of the general scope to this special case. Hence, in spite of the transversality of the field, the operator π could have non-vanishing matrix elements $\langle\lambda'|\,\pi\,|\lambda\rangle$ with $\lambda, \lambda' = 0$ in case where the direction of propagation $\boldsymbol{e}$ of the pumping light is not along the z-direction of the atomic triad $\boldsymbol{e}_i$. Actually, throughout in this context we treat polarization rather by a 3×3-matrix than by a 2×2-matrix as could be done, if we related polarization to the direction of the radiation field only.

We further introduce the operator $\mathscr{A}(\lambda, \lambda')$:

$$\langle a, \alpha|\,\mathscr{A}(\lambda, \lambda')\,|a, \alpha'\rangle = |\langle b\,\|\,D^{(1)}\,\|\,a\rangle|^{-2} \sum_{\beta} \langle a, \alpha|\,D_{\lambda'}^{(1)+}\,|b, \beta\rangle \cdot \langle b, \beta|\,D_{\lambda}^{(1)}\,|a, \alpha'\rangle \quad (35)$$

which, according to (30), comprises a summation over products of Clebsch-Gordan coefficients involved in the pumping process and for which holds the conjugation relation $\mathscr{A}^{+}(\lambda, \lambda') = \mathscr{A}(\lambda', \lambda)$.

Using the definitions for π and $\mathscr{A}(\lambda, \lambda')$ together with the absorption term (25b) we finally obtain from (11b):

$$\langle a, \alpha|\,\dot{\mu}_{\mathrm{at}}\,|b, \beta\rangle_{\mathrm{abs.}} = -\big((2T_p)^{-1} - i\hbar^{-1}\,\varDelta E_p\big)\,\hat{d}^2 \sum_{\substack{\alpha'\\ \lambda,\lambda'}} \langle a, \alpha|\,\mathscr{A}(\lambda, \lambda)\,|a, \alpha'\rangle\,\langle\lambda'|\,\pi\,|\lambda\rangle \times \langle a, \alpha'|\,\mu_{\mathrm{at}}\,|b, \beta\rangle \quad (36)$$

with

$$1/T_p = \Gamma_p = \big(2\hbar(2\pi)^2\,\varepsilon_0\,\hat{d}^2\big)^{-1}\,|\langle b\|\,D^{(1)}\,\|a\rangle|^2\,\omega_0^2 \int k^2\,dk\langle n(\omega_k)\rangle\,\omega_k^{-1}\,\delta(\omega_0 - \omega_k) \quad (37)$$

and

$$\varDelta E_p = \big(2(2\pi)^3\,\varepsilon_0\hat{d}^2\big)^{-1}\,|\langle b\|\,D^{(1)}\,\|a\rangle|^2\,\omega_0^2\,\mathrm{PP}\!\int k^2\,dk\langle n(\omega_k)\rangle\,\omega_k^{-1}(\omega_0 - \omega_k)^{-1},$$

accepting a δ-distribution for the Ω-dependence of the spectral function $\langle n(\omega_k)\rangle$ around the beam direction $\boldsymbol{e}$. T_p is the mean life time (pumping time constant) and $\varDelta E_p$ the level shift of the atomic ground state under the influence of the pumping light beam.

Here again we ignored a possible level spacing of the magnetic sublevels in the ground and excited state as might be induced by interaction with an external magnetic field. The arguments used in treating the spontaneous emission term can be repeated here, but the behaviour of the spectral function $\langle n(\omega_k)\rangle$ over the range of absorption frequencies will introduce further difficulties. Dealing with a single pumping time constant T_p and level shift $\varDelta E_p$ is correct only in cases, where $\langle n(\omega_k)\rangle$ covers the different absorption lines so that there are no appreciable changes of this function over the range of absorption (same T_p) and that the width $\varDelta\omega_0$ of the absorption range is small compared with the line width $\varDelta\omega$ of the pumping light beam (same $\varDelta E_p$). These conditions, in general, are preserved only for optical pumping in weak external magnetic fields limiting the useful factorized separation of intensity (T_p, $\varDelta E_p$) and spatial symmetry problems $\big(\pi, \mathscr{A}(\lambda', \lambda)\big)$ to the weak magnetic field case. This simplification, being not essential for the general formulation of the theory [*6*], constitutes a fundamental condition for the advantages of spherical tensor notation.

Kinetic equations for $\mu(a)$ and $\mu(b)$:

Independent of these considerations, equations (25) or more obviously (31, 36) show that for the stationary case the matrix elements of μ_{at} connecting the ground and excited states vanish. They are solutions of a system of homogeneous linear differential equations and hence remain zero if they were zero initially or decay with time constants which are in the order of T_s. Therefore, at any time μ_{at} can be written as a direct sum of two operators $\mu(a)$ and $\mu(b)$ which have matrix elements between ground states (index a) or excited states (index b) only:

$$\mu_{\text{at}} = \mu(a) + \mu(b). \tag{38}$$

Let P_a and P_b be the traces of $\mu(a)$ and $\mu(b)$:

$$P_a = \operatorname{tr} \mu(a), \qquad P_b = \operatorname{tr} \mu(b). \tag{39}$$

Then, the normalization condition on μ_{at} requires:

$$P_a + P_b = 1, \tag{40}$$

whence P_a and P_b can be interpreted as the probabilities of finding an atom in the ground or excited state respectively and for the statistical operators $\varrho(a)$ and $\varrho(b)$ we have:

$$\varrho(a) = P_a^{-1}\mu(a), \qquad \varrho(b) = P_b^{-1}\mu(b). \tag{41}$$

This reduction of the full system (11b) of differential equations for the matrix elements of μ_{at} is due to the assumption of a conventional light source with thermal-like statistics stated above.

Calculating the time derivative of the remaining matrix elements $\langle b, \beta| \mu_{\text{at}} |b, \beta'\rangle$ and $\langle a, \alpha| \mu_{\text{at}} |a, \alpha'\rangle$ in a quite analogous manner we finally obtain:

$$\begin{aligned}\langle b, \beta| \dot{\mu}_{\text{at}} |b, \beta'\rangle = &-T_s^{-1}\langle b, \beta| \mu_{\text{at}} |b, \beta'\rangle + \hat{a}^2(T_p\hat{1}^2)^{-1} \sum_{\substack{\lambda,\lambda' \\ \alpha,\alpha'}} \langle\lambda| \pi |\lambda'\rangle \langle a, \alpha| \mu_{\text{at}} |a, \alpha'\rangle \\ &\times (-1)^{2a-(\alpha+\alpha')} \langle b, \beta; a, -\alpha \mid 1, \lambda\rangle \langle b, \beta'; a, -\alpha' \mid 1, \lambda'\rangle\end{aligned} \tag{42}$$

and

$$\begin{aligned}\langle a, \alpha| \dot{\mu}_{\text{at}} |a, \alpha'\rangle = T_s^{-1} &\sum_{\beta,\beta'} \delta_{\beta'\beta,\alpha-\alpha'} \langle 1, \beta - \alpha; a, \alpha \mid b, \beta\rangle \langle b, \beta| \mu_{\text{at}} |b, \beta'\rangle \\ &\times \langle 1, \beta' - \alpha'; a, \alpha' \mid b, \beta'\rangle - \sum_{\lambda,\lambda'} \langle\lambda| \pi |\lambda'\rangle \hat{a}^2\{\langle a, \alpha| \mathscr{A}(\lambda, \lambda') \mu_{\text{at}} |a, \alpha'\rangle \\ &\times [(2T_p)^{-1} - i\hbar^{-1} \Delta E_p] + \langle a, \alpha| \mu_{\text{at}} \mathscr{A}(\lambda, \lambda') |a, \alpha'\rangle \\ &\times [(2T_p)^{-1} + i\hbar^{-1} \Delta E_p]\}.\end{aligned} \tag{43}$$

These are the kinetic equations of Cohen-Tannoudji written in a form which holds for any direction of propagation and any statistical mixture of polarization states of the incident pumping light beam. Using the technique outlined in the appendix we take these equations now as a starting point for the introduction of spherical tensor notation.

3.3. The spherical tensor notation

The spherical tensor notation of the kinetic equations follows by straightforward calculation from the transformation formula (A10) between the matrix elements and the spherical tensor operator components of an operator. Differentiating (A10) with respect to time and inserting the time derivatives of the matrix elements of $\mu(j)$ according to (42, 43) we have:

$$\dot{\mu}_{K,\varkappa}(b) = -T_s^{-1}\mu_{K,\varkappa}(b) + T_p^{-1} \sum_{\substack{K_a,\varkappa_a \\ L,\lambda}} \mu_{K_a,\varkappa_a}(a)\, \pi_{L,\lambda}\hat{a}^2\hat{K}_a\hat{L}\langle K_a, \varkappa_a;\, L, \lambda \mid K, \varkappa\rangle \begin{Bmatrix} a & 1 & b \\ a & 1 & b \\ K_a & L & K \end{Bmatrix} \tag{44}$$

and

$$\dot{\mu}_{K,\varkappa}(a) = -(-1)^{a-b+1+K} \sum_{\substack{K_a,\varkappa_a\\L,\lambda}} \mu_{K_a,\varkappa_a}(a)\, \pi_{L,\lambda}\, \hat{a}^2 \hat{K}_a \hat{L}\, \langle K_a, \varkappa_a; L, \lambda \mid K, \varkappa\rangle \begin{Bmatrix} L & 1 & 1 \\ b & a & a \end{Bmatrix} \begin{Bmatrix} K_a & L & K \\ a & a & a \end{Bmatrix}$$
$$\times \left[\left((2T_p)^{-1} + i\hbar^{-1}\, \Delta E_p\right) + \left((2T_p)^{-1} - i\hbar^{-1}\, \Delta E_p\right) \cdot (-1)^{K_a+L+K}\right]$$
$$+ T_s^{-1} \mu_{K,\varkappa}(b)\, \hat{b}^2 \hat{K} \hat{1} \begin{Bmatrix} a & 1 & b \\ a & 1 & b \\ K & 0 & K \end{Bmatrix}. \tag{45}$$

These are the kinetic equations for the spherical tensor operator components of $\mu(a)$ and $\mu(b)$. They constitute a system of coupled differential equations of the first order whose coefficients are mainly determined by the radiation parameters T_s, T_p, ΔE_p (intensity) and $\pi_{L,\lambda}$ (polarization). The occuring vector coupling coefficients up to the order of $9j$-symbols express the conservation of angular momentum in the elementary process of absorption and emission. For these coefficients we used the definitions and conventions of MESSIAH [10].

We make some brief comments in order to illustrate the advantages of this notation (see also [3]).

To what degree of coupling the interdependence of the tensor components arises depends on the polarization of the pumping light beam. Since, by definition, π is an operator which acts in a state space with total angular momentum quantum numer $j = 1$, the tensor order L (L integer) of $\pi_{L,\lambda}$ is limited to $L \leqq 2$. If we consider an unpolarized pumping light beam we have for non-vanishing $\pi_{L,\lambda}$ L even and the symmetry relations for the special $9j$-symbol in (44) imply for non-vanishing contributions $K_a + K$ to be even too. So the system (44) decouples into two systems, one for K, K_a even and the other for K, K_a odd. Caused by the factor $(-1)^{K_a+L+K}$ the same holds for the real transition terms in (45) and in cases, where the 'virtual' transition terms can be ignored ($\Delta E_p \ll \hbar T_p^{-1}$), only the system with K, K_a even has a non-trivial stationary solution. In practice, such an a priori reduction could establish a welcome facilitation in dealing with kinetic equations.

As for the 'virtual' transition terms in (45), which are proportional to ΔE_p, a comparison with the kinetic equations due to magnetic interaction (Sec. 3.4.) shows that there might be a strict formal analogy depending on the polarization and direction of propagation of an off-resonant optical radiation field. It is this analogy that led to the often used characterization of these tensor light shifts as 'fictitious magnetic field' effects [16, 17].

The $\mu_{K,\varkappa}(j)$ ($j = a, b$) are proportional to the state multipoles $\varrho_{K,\varkappa}(j)$ via the probability factors P_j according to (41). Since $\varrho_{0,0}(j)$ is a mere normalization factor, the system (44, 45) for $K = 0$ gives the differential equations for the, generally, time dependent probabilities P_a and P_b of finding an atom in the ground or excited state respectively. The significance of these probabilities will be discussed further in Sec. 3.5.

3.4. External magnetic fields

We now take into account the interaction with an external magnetic field whose decomposition with respect to the original frame might be given in the form:

$$\boldsymbol{H} = H_x \boldsymbol{e}_x + H_y \boldsymbol{e}_y + H_z \boldsymbol{e}_z \tag{46}$$

and which is assumed to be the same for all the atoms in question. If W is the interaction Hamiltonian, the Liouville equation becomes:

$$i\hbar \dot{\mu}_{\mathrm{at}} = [W, \mu_{\mathrm{at}}]. \tag{47}$$

Treating the interaction as a weak perturbation the operator W decomposes into a direct sum:

$$W = W(a) + W(b)$$

and for the interaction Hamiltonian $W(j)$ we use the semi-classical form (for a strict quantum-electrodynamic treatment see for instance [*18*]):

$$\begin{aligned} W(j) &= -g_j\mu_B \boldsymbol{H}\cdot\boldsymbol{J} \\ &= -\hbar\big(h_0(j)\,J_z + h_+(j)\,J_+ + h_-(j)\,J_-\big) \end{aligned} \tag{48}$$

with

$$h_0(j) = g_j\hbar^{-1}\mu_B H_z, \qquad h_\pm(j) = g_j(2\hbar)^{-1}\,\mu_B(H_x \mp iH_y) \qquad j = a, b.$$

Here, g_j is the g-factor of the atomic state $|j, m\rangle$ and μ_B the Bohr magneton.
Inserting this expression for W and introduction the irreducible tensor expansion of $\mu(j)$ in (47) we can replace the commutators by the commutating relations (A2) for the spherical tensor operators and with the trace orthogonality (A5) we find the kinetic equations for $\mu_{K,\varkappa}(j)$ due to magnetic interaction:

$$\begin{aligned} \dot\mu_{K,\varkappa}(j) = -i\big[&\varkappa h_0(j)\,\mu_{K,\varkappa}(j) \\ &+ h_+(j)\,\big(K(K+1) - \varkappa(\varkappa+1)\big)^{1/2}\,\mu_{K,\varkappa+1}(j) + \\ &+ h_-(j)\,\big(K(K+1) - \varkappa(\varkappa-1)\big)^{1/2}\,\mu_{K,\varkappa-1}(j)\big]. \end{aligned} \tag{49}$$

The dynamics of the atomic system is given by an incoherent superposition of the optical pumping terms (44, 45) and the magnetic interaction terms (49) and should be completed by appropriate relaxation terms due to the interaction with a thermal bath [*3, 19, 20, 21*][5]).
Eqs. (49) mix the kinetics of the irreducible tensor components of one given order K. Usually, in magnetic resonance experiments the coefficients $h(j)$ are time dependent and so the kinetic equations form a system of coupled linear differential equations of the first order with time dependent coefficients. There are only a few experimental situations for which at least the stationary solutions of this system can be given in a closed analytic form. Considerable effort had been undertaken in order to develop several methods of approximation [*5, 22, 25*] and here again the irreducible tensor notation, which probably leads to an a priori reduction of the entire system of differential equations for the tensor components of μ_{at}, constitutes an efficient tool in handling with this problem. Further interest in this topics may come from recent advances in nonlinear optics, where a number of coherent optical effects have their counterparts in magnetic resonance [*26*].

3.5. Reduction to the ground state kinetics

One of the most important approximations in handling with kinetic equations for the atomic system starts with the remark that the mean life time T_s is the leading time constant of the excited state kinetics. Other interaction time constants are the thermal relaxation time constant $\Theta_{K,\varkappa}(b)$, the period of the Larmor precession in an external magnetic field or else the period $2\pi/\omega$ of the rf-part of this field when studying magnetic resonance effects and perhaps a characteristic period $2\pi/\omega_p$ introduced by intensity modulation of the pumping light beam governing the time dependence of the pumping

[5]) The interaction with a thermal bath in the presence of a static external magnetic field H would introduce a polarization of the atomic ground state (Boltzmann distribution). We always assume this effect to be small compared with the orientation achieved by optical pumping.

time constant T_p. Furthermore we had to consider the influence of the ground state kinetics governed by essentially the same or analogous time constants.
If we take the conditions for optical pumping the odd isotopes of mercury as a guide, we can assume:

$$T_s^{-1} \gg T_p^{-1},\ \Theta_{K,\varkappa}(b),\ \omega,\ \omega_p \tag{50}$$

excluding magnetic resonances in the excited state and intensity modulation of the pumping light with frequencies in the order of T_s^{-1}. This is essentially the extended secular approximation of Cohen-Tannoudji which means that we can take the stationary solutions of the cmbined equations (44, 49) in the form:

$$\begin{aligned} &\left(T_s^{-1} + i\varkappa h_0(b)\right) \mu_{K,\varkappa}(b) \\ &+ ih_+(b) \left(K(K+1) - \varkappa(\varkappa+1)\right)^{1/2} \mu_{K,\varkappa+1}(b) \\ &+ ih_-(b) \left(K(K+1) - \varkappa(\varkappa-1)\right)^{1/2} \mu_{K,\varkappa-1}(b) \\ = &\ T_p^{-1} F_{K,\varkappa}(t) \end{aligned} \tag{51}$$

with

$$F_{K,\varkappa}(t) = \hat{a}^2 \sum_{\substack{K_a,\varkappa_a \\ L,\lambda}} \mu_{K_a,\varkappa_a} \pi_{L,\lambda} \hat{K}_a \hat{L} \langle K_a, \varkappa_a; L, \lambda \mid K, \varkappa \rangle \begin{Bmatrix} a & 1 & b \\ a & 1 & b \\ K_a & L & K \end{Bmatrix}$$

to describe the excited state kinetics. Thus the system of differential equations for the excited state is reduced to a system of algebraic equations in the tensor components $\mu_{K,\varkappa}(b)$. We use to refer to these conditions as to the algebraic approximation for the excited state kinetics.
Inserting these solutions into the combined equations (45, 49) we are left with a system of differential equations for the ground state multipoles alone. Then, the kinetic equations for the atomic ground state multipoles are fundamental for the description of the whole system. The ground state solutions determine the kinetics of the excited state via (51) and in the following sections we will present a model for the calculation of optical signal functions in terms of the atomic state multipoles.
Before writing down explicit expressions for the kinetic equations of the ground state we make some remarks on the probabilities P_a and P_b of finding an atom in the ground or excited state respectively.
As mentioned above, equation (51) written down for $K = 0$ gives a solution for the excitation probability P_b. Together with (40, 41, A4) we have:

$$P_b(t) = T_s T_p^{-1} \hat{b} F_{0,0}(t) \left(1 + T_s T_p^{-1} \hat{b} F_{0,0}(t)\right)^{-1}. \tag{52}$$

The dominant factor in (52) is the ratio $T_s T_p^{-1}$ which under usual experimental conditions is of the order of 10^{-5} or less. The function $F_{0,0}(t)$ is bounded since it is a finite sum over products of irreducible tensor components of statistical operators. To get an idea of the order of magnitude of $F_{0,0}(t)$ consider an isotrope unpolarized aillumination of the ensemble of atoms. In that case π is proportional to the unity matrix whence $\hat{b} F_{0,0}(t) = 1$. This consideration suggests the following approximate expressions for P_a and P_b:

$$P_a = 1, \qquad P_b = T_s T_p^{-1} \hat{b} F_{0,0}(t) \ll 1. \tag{53}$$

We now return to the algebraic approximation (51) for the excited state kinetics. A considerable simplification is stated by the assumption that beside (50) the inequality

$$T_s^{-1} \gg h_\pm(b) \tag{54}$$

holds too. In that case the stationary solutions have the form:

$$\mu_{K,\varkappa}(b) = T_s T_p^{-1} F_{K,\varkappa}(t) \left(1 + i\varkappa h_0(b)\, T_s\right)^{-1} \tag{55}$$

and we use these expressions in order to calculate the kinetic equations for the ground state multipoles.

It should be noted that certainly the approximate solutions (55) are sufficient for the derivation of kinetic equations for the ground state but care should be taken when calculating the excited state multipoles in terms of the ground state parameters, especially in such cases, where $F_{K,\varkappa}(t)$ vanishes identically. For that purpose equations (51) are the better ones.

Inerting (55) into the ground state equations we obtain:

$$\begin{aligned}\dot{\varrho}_{K,\varkappa}(a) = &-i\left[\varkappa h_0(a)\, \varrho_{K,\varkappa}(a) + h_+(a) \left(K(K+1) - \varkappa(\varkappa+1)\right)^{1/2}\right.\\ &\left.\times\, \varrho_{K,\varkappa+1}(a) + h_-(a) \left(K(K+1) - \varkappa(\varkappa-1)\right)^{1/2} \varrho_{K,\varkappa-1}(a)\right]\\ &+ \hat{a}^2 (2T_p)^{-1} (-1)^{a+b+K+1} \sum_{\substack{K_a,\varkappa_a\\ L,\lambda}} \varrho_{K_a,\varkappa_a}(a)\, \pi_{L,\lambda} \hat{K}_a \hat{L} \cdot \langle K_a, \varkappa_a; L, \lambda \mid K, \varkappa\rangle\\ &\times\, \hat{C}_{K,\varkappa}(a, b, K_a, L) - \sum_{K_a,\varkappa_a} (T^{K,\varkappa}_{K_a,\varkappa_a})^{-1}\, \varrho_{K_a,\varkappa_a}(a)\end{aligned} \tag{56}$$

with

$$\begin{aligned}\hat{C}_{K,\varkappa}(a, b, K_a, L) = &\, 2\hat{b}^2 \left(1 + i\varkappa h_0(b)\, T_s\right)^{-1} \begin{Bmatrix} a & 1 & b\\ a & 1 & b\\ K_a & L & K \end{Bmatrix} \begin{Bmatrix} K & b & b\\ 1 & a & a \end{Bmatrix}\\ &- (-1)^{2a} \begin{Bmatrix} L & 1 & 1\\ b & a & a \end{Bmatrix} \begin{Bmatrix} K_a & L & K\\ a & a & a \end{Bmatrix}\\ &\times \left[(1 + i2T_p \hbar^{-1} \Delta E_p) + (1 - i2T_p \hbar^{-1} \Delta E_p)\, (-1)^{K_a+L+K}\right].\end{aligned}$$

For this notation we used (53) in order to leave the distinction between the irreducible tensor operator components of $\mu(a)$ and the state multipoles $\varrho_{K,\varkappa}(a)$. Eq. (53) shows, that this distinction being not essential for the ground state is important for the formulation of the excited state kinetics.

The last term in (56) is added to give a formal description of thermal relaxation [*3*, *21*]. The coefficients $\hat{C}_{K,\varkappa}$ comprise a summation over vector coupling coefficients. In cases, where the level shift ΔE_p can be neglected, we can write:

$$\begin{aligned}\hat{C}_{K,\varkappa} = &\left(1 + (-1)^{K_a+L+K}\right)\\ &\times \left(\hat{b}^2 \left(1 + i\varkappa h_0(b)\, T_s\right)^{-1} \begin{Bmatrix} a & 1 & b\\ a & 1 & b\\ K_a & L & K \end{Bmatrix} \begin{Bmatrix} K & b & b\\ 1 & a & a \end{Bmatrix} - (-1)^{2a} \begin{Bmatrix} L & 1 & 1\\ b & a & a \end{Bmatrix} \begin{Bmatrix} K_a & L & K\\ a & a & a \end{Bmatrix}\right),\end{aligned} \tag{57}$$

where we used symmetry relations of the $9j$-symbol.

Eqs. (56) together with the special form (57) for the $\hat{C}_{K,\varkappa}$ are essential the equations (6) of reference [*4*].

4. Rate equations for spontaneous emission and absorption

Till now we only used that part of the Liouville equation (11) that gives the kinetics of the atomic system in the pumping region. We achieved decoupling from the kinetics of the radiation field by the assumption that μ_{rad} is a given constant determined by the incident pumping light beam alone. Under some aspects this model might appear quite one-sided. In the same manner we could ask for the rate of change in the statistical operator for the radiation field interacting with a given ensemble of atoms described by μ_{at}. In this way, with the resulting modification in μ_{rad} as a function of the spatial variables for instance, we could obtain a second order correction for μ_{at} along the direction of the pumping light beam propagating in the atomic vapour.
In order to obtain stationary solutions, however, we had to specify in any way the coupling of the radiation to the light source and the interaction with the walls of the pumping region. Different approaches to this problem are described elsewhere [*6*, *9*]. In our context and for the discussion of optical pumping signals in dilute gases it is sufficient to postulate the existence of such stationary solutions and to look at the rate of change in μ_{rad}, only. The rate of change in μ_{rad} due to spontaneous emission gives the observed fluorescence radiation and the rate of change due to absorption is proportional to the signal observed in transmission experiments.
The calculation follows strictly the scheme outlined in Sec. 3. We make the same assumptions concerning the statistics of the radiation field and as before we restrict the representation to the utmost one-quantum states of the different field modes. Starting with the Liouville equation in the form:

$$\dot{\mu}_{\text{rad}} = -(\eta/\hbar)^2 N \operatorname{tr}_{\text{at}}(V\mathcal{L}\mu - V\mu\mathcal{L} - \mathcal{L}\mu V + \mu\mathcal{L}V) \tag{58}$$

with

$$\mu = \mu_{\text{at}}\mu_{\text{rad}}$$

we demonstrate the different steps guided by the first term $\operatorname{tr}_{\text{at}} V\mathcal{L}\mu$. Denoting the general state in the Fock space of the field by $|r\rangle$ we have for the matrix element $\langle r| \operatorname{tr}_{\text{at}} V\mu |r'\rangle$:

$$\begin{aligned}
&\langle r| \operatorname{tr}_{\text{at}} V\mu |r'\rangle \\
&= \sum_{\substack{\alpha,\alpha',\beta \\ \boldsymbol{k},\boldsymbol{k}' \\ \sigma,\sigma'}} \langle r|\, a^{+}_{\boldsymbol{k},\sigma} a_{\boldsymbol{k}',\sigma'} \mu^{\text{r}}{}_{\text{ad}}\, |r'\rangle \langle a, \alpha|\, V^{+}_{\boldsymbol{k},\sigma}\, |b, \beta\rangle \\
&\quad \times \langle b, \beta|\, V_{\boldsymbol{k}',\sigma'}\, |a, \alpha'\rangle \langle a, \alpha'|\, \mu_{\text{at}}\, |a, \alpha\rangle\; 2\pi\delta_{-}(\omega_0{}^{(\beta,\alpha')} - \omega_{k'}) \\
&\quad + \sum_{\substack{\beta,\beta',\alpha \\ \boldsymbol{k},\boldsymbol{k}' \\ \sigma,\sigma'}} \langle r|\, a_{\boldsymbol{k},\sigma} a_{\boldsymbol{k}',\sigma'} \mu_{\text{rad}}\, |r'\rangle \langle b, \beta|\, V_{\boldsymbol{k},\sigma}\, | a, \alpha\rangle \\
&\quad \times \langle a, \alpha|\, V^{+}_{\boldsymbol{k}',\sigma'}\, |b, \beta'\rangle \langle b, \beta'|\, \mu_{\text{at}}\, |b, \beta\rangle\; 2\pi\delta_{-}(\omega_{k'} - \omega_0{}^{(\beta',\alpha)}).
\end{aligned} \tag{59}$$

As a consequence of (38) the matrix element (59) like the other ones, decomposes into two parts, one of which is related to the atomic ground state and thus to the absorption process whereas the other one represents emission. Since in our model, we neglect induced emission as well as the reabsorption of a once spontaneously emitted photon within the pumping region, emission and absorption are completely independent.
If we were interested in the kinetics of one single mode group, say $\boldsymbol{k}_0$, with the direction of propagation $\boldsymbol{e}$, we had to take the partial trace over the different and independent mode groups of the field. The expectation values $\langle a^{+}_{\boldsymbol{k},\sigma}\rangle$ and $\langle a_{\boldsymbol{k},\sigma}\rangle$ vanish and the summation

over the mode variables in (59) reduces to a summation over the polarization parameters of the mode group $\boldsymbol{k}_0$.

The dependence upon the atomic variables can be summarized in an efficient way, if we introduce two auxiliary operators according to the definitions:

$$\langle 1, \lambda|\, \varrho(A)\, |1, \lambda'\rangle = \hat{a}^2 \hat{1}^{-2} \sum_{\alpha,\alpha',\beta} \langle b, \beta; a, -\alpha \mid 1, \lambda'\rangle \times (-1)^{a-\alpha} \langle a, \alpha|\, \varrho(a)\, |a, \alpha'\rangle\, (-1)^{a-\alpha'} \langle b, \beta; a, -\alpha' \mid 1, \lambda\rangle \quad \text{(absorption)} \tag{60}$$

$$\langle 1, \lambda|\, \varrho(S)\, 1, \lambda'\rangle = \hat{b}^2 \hat{1}^{-2} \sum_{\alpha,\beta,\beta'} \langle b, \beta; a, -\alpha \mid 1, \lambda\rangle \times (-1)^{a-\alpha} \langle b, \beta|\, \varrho(b)\, |b, \beta'\rangle\, (-1)^{a-\alpha} \langle b, \beta'; a - \alpha \mid 1, \lambda'\rangle \quad \text{(emission)}.$$

We call these operators the atomic absorption and emission operators, respectively. $\varrho(A)$ and $\varrho(S)$ have all the properties which are essential for statistical operators and we can look upon them as statistical operators for electric dipole transitions within the atomic system[6]).

With these definitions, equation (59) written down for a single mode group becomes:

$$\langle r_0|\, \mathrm{tr}_{\mathrm{at}} V_{\boldsymbol{k}_0} \mathscr{L}_{\boldsymbol{k}_0} \mu_{\boldsymbol{k}_0}\, |r_0'\rangle = \sum_{\substack{\sigma,\sigma' \\ \lambda,\lambda'}} [R^{(1)*}_{\lambda,\sigma}(\boldsymbol{e})\, \langle r_0|\, a^+_{\boldsymbol{k}_0,\sigma} a_{\boldsymbol{k}_0,\sigma'} \mu_{\boldsymbol{k}_0}\, |r_0'\rangle\, R^{(1)}_{\lambda',\sigma'}(\boldsymbol{e}) \times P_a \hat{a}^{-2} \langle 1, \lambda|\, \varrho(A)\, |1, \lambda'\rangle\, f^*(\omega_{\boldsymbol{k}0}) + R^{(1)}_{\lambda,\sigma}(\boldsymbol{e}) \times \langle r_0|\, a_{\boldsymbol{k}_0,\sigma} a^+_{\boldsymbol{k}_0,\sigma'} \mu_{\boldsymbol{k}_0}\, |r_0'\rangle\, R^{(1)*}_{\lambda',\sigma'}(\boldsymbol{e})\, P_b \hat{b}^{-2} \langle 1, \lambda'|\, \varrho(S)\, |1, \lambda\rangle\, f(\omega_{\boldsymbol{k}_0})] \tag{61}$$

with

$$f(\omega) = \omega_0{}^2 \omega^{-1}\, |\langle a\, \|D^{(1)}\|\, b\rangle|^2\, 2\pi \delta_-(\omega - \omega_0).$$

Similar equations are obtained for the remaining three other terms in (58).

Collecting all these terms the sequence of the operators a, a^+ and μ is important. We give a representation of the various operator products in the space of the utmost one-quantum states using a slightly simplified notation and neglecting $\langle n\rangle$ with respect to 1 (cf. Sec. 3.1):

$$\begin{aligned} \mu a_\sigma a^+_{\sigma'} &= a_\sigma a^+_{\sigma'} \mu = |\mathrm{vac}\rangle\, \delta_{\sigma,\sigma'}\, \langle \mathrm{vac}| \\ \mu a_\sigma{}^+ a_{\sigma'} &= \sum_{\sigma''} |\sigma''\rangle\, \langle n\rangle\, \langle \sigma''|\, \pi\, |\sigma\rangle\, \langle \sigma'| = (a^+_{\sigma'} a_\sigma \mu)^+ \\ a_\sigma{}^+ \mu a_{\sigma'} &= |\sigma\rangle\, \langle \sigma'| \\ a_\sigma \mu a^+_{\sigma'} &= |\mathrm{vac}\rangle\, \langle n\rangle\, \langle \sigma|\, \pi\, |\sigma'\rangle\, \langle \mathrm{vac}|. \end{aligned} \tag{62}$$

We are then left with four rate equations, two for absorption and emission, respectively. The components of each pair of rate equations are not mutually independent but are

[6]) The introduction of these operators is convenient in case of a pure multipole radiation of one given order K without specifying a certain initial or final substate for the involved atomic transition. The rate equations (67, 68) can then be formulated by means of the expectation value of the polarization operator π with these statistical operators.
If we wanted to study the emission or absorption from one given initial atomic substate, we had to redefine $\varrho(b)$ or $\varrho(a)$ in (60) as to be the projector onto this substate. A further restriction to a specific final atomic substate then imposes limitations upon the possible polarization operators that occur in the rate equations.

related by the conservation of trace. Thus the equation:

$$\langle 0, \boldsymbol{k}_0 | \dot{\mu}_{\boldsymbol{k}_0} | 0, \boldsymbol{k}_0 \rangle = - \sum_{\sigma} \langle 1, \boldsymbol{k}_0, \sigma | \dot{\mu}_{\boldsymbol{k}_0} | 1, \boldsymbol{k}_0, \sigma \rangle \tag{63}$$

holds for both, absorption and emission.
Considering a general polarization state of the field with wave vector $\boldsymbol{k}_0$:

$$|1, \boldsymbol{k}_0, \varepsilon \rangle = \sum_{\sigma} |1, \boldsymbol{k}_0, \sigma \rangle \langle \sigma \mid \varepsilon \rangle, \tag{64}$$

the results can be given in terms of the rate of change in the average photon number:

$$\langle n_{\boldsymbol{k}_0, \varepsilon} \rangle = \mathrm{tr}\,(a^{+}_{\boldsymbol{k}_0, \varepsilon} a_{\boldsymbol{k}_0, \varepsilon} \mu_{\boldsymbol{k}_0}) = \sum_{\sigma, \sigma'} \langle \sigma \mid \varepsilon \rangle \langle \varepsilon \mid \sigma' \rangle \langle \sigma' | \mu_{\boldsymbol{k}_0} | \sigma \rangle. \tag{65}$$

But since we are interested in the total radiation in the direction of observation $\boldsymbol{e}$ rather than in the kinetics of one single mode group we write for the photon number operator:

$$n_\varepsilon = \sum_{(\boldsymbol{k}/k) = \boldsymbol{e}} a^{+}_{\boldsymbol{k}, \varepsilon} a_{\boldsymbol{k}, \varepsilon}$$

and for the statistical operator of this polarization state we have:

$$\langle \sigma | \pi_{\boldsymbol{e}}(\varepsilon) | \sigma' \rangle = \langle \sigma \mid \varepsilon \rangle \langle \varepsilon \mid \sigma' \rangle. \tag{66}$$

Performing the summation over $\boldsymbol{k}$ according to the scheme (26) we finally obtain the rate equations for spontaneous emission and absorption in the form:

$$\langle \dot{n}_\varepsilon \rangle_{\mathrm{em.}} = (3/2)\, N P_b T_s^{-1}\, (d\Omega/4\pi)\, \mathrm{tr}\, \big(\varrho(S)\, \pi(\varepsilon)\big), \tag{67}$$

$$\langle \dot{n}_\varepsilon \rangle_{\mathrm{abs.}} = -N P_a T_p'^{-1}\, \mathrm{tr}\, \big(\varrho(A)\, \hat{\pi}(\varepsilon)\big) \tag{68}$$

with

$$\hat{\pi}(\varepsilon) = (1/2)\, [\pi', \pi(\varepsilon)]_+ - i\hbar^{-1} T_p'\, \Delta E_p' [\pi', \pi(\varepsilon)],$$

where N is the number of participant atoms.
In the *emission rate equation* (67), P_b is the probability of finding an atom in the excited state (cf. (39, 52)), T_s the mean life time of this atomic state and $\varrho(S)$ the atomic emission operator (60). In the direction of observation $\boldsymbol{e}$, the polarization ε is described by the polarization operator $\pi_{\boldsymbol{e}}(\varepsilon)$ according to (64, 66). If we relate polarization rather to the atomic frame than to the direction of observation $\boldsymbol{e}$, the polarization operator $\pi_{\boldsymbol{e}}(\varepsilon)$ transforms according to (34), where the rotation $\mathscr{R}(\boldsymbol{e})$ transforms the cartesian triad $\boldsymbol{e}_i (i = 1, 2, 3)$, used for the description of atomic states, the atomic frame, into the triad $\boldsymbol{e}_i(\boldsymbol{k})$, which is the basis for the description of signal radiation, the signal frame, and for which $\boldsymbol{e}_3(\boldsymbol{k}) = \boldsymbol{e}$ (cf. Sec. 5.4). It is the transformed polarization matrix $\pi(\varepsilon)$ that occurs in the notation of (67, 68)[7].
In the *absorption rate equation* (68) we use a prime in order to distinguish the radiation parameters T_p', $\Delta E_p'$ and the transformed polarization operator π' of the sample beam from those for the main pumping light beam. In transmission experiments the sample beam for observation of the atomic ground state multipoles might be different from the main pumping beam. Furthermore, the distinction between the polarization operators π'

[7]) If we relate the representation of the polarization operator to the atomic frame, we leave the index $\boldsymbol{e}$

and $\pi(\varepsilon)$ accounts for a possible difference between the polarization π' of the sample beam and the polarization $\pi(\varepsilon)$ of the detector used in the signal channel[8]).
The form of the hermitean operator $\hat{\pi}(\varepsilon)$ in (68) accounts for a structural difference in the emission and absorption rate equations. $\hat{\pi}(\varepsilon)$ is a function in the polarization operators π' and $\pi(\varepsilon)$. The anticommutator ($[,]_+$) and the commutator terms in (68) are related to the so called 'real' and 'virtual' transitions, respectively [*2*]. There are a number of important experiments, for which the virtual transition term is significant [*2*, *27*].
The special form of the rate equations illustrate the advantage of the definitions (60) for the atomic emission and absorption operators. The emission and absorption rate can be given in terms of expectation values of the polarization operator or simple functions of those operators taken with the statistical operators $\varrho(S)$ and $\varrho(A)$. Let us take a glance at the emission term (67) which is comparatively simple.
The emission rate of photons with polarization ε is proportional to the expectation value of $\pi(\varepsilon)$ taken with $\varrho(S)$. If we ask for the rate of photons without looking neither at the direction of emission nor at the polarization state, for instance, we introduce the operator

$$\pi_e(\text{unpol.}) = \sum_\sigma \pi_e(\sigma)$$

for which we obtain according to (34):

$$\langle\lambda|\ \pi\ (\text{unpol.})\ |\lambda'\rangle = \sum_\sigma R^{(1)}_{\lambda,\sigma}(e)\ R^{(1)*}_{\lambda',\sigma}(e)\,.$$

Integrating over all directions and using (33) the total emission rate is found to be:

$$\langle\dot{n}\rangle_{\text{em.,tot.}} = NP_bT_s^{-1}.$$

Similar illustrations could be given for the absorption rate equation (68). The differences arise from the fact, that unlike the spontaneous emission case, where the interaction couples the atomic states to the isotropic photon vacuum, the absorption process is mainly determined by the unisotropic structure and the photon distribution of the light beam. However, if the polarization operator $\pi(\varepsilon)$ of the detection channel commutes with the polarization operator π' of the sample beam, the ground state level shift due to interaction with this beam cannot be observed. In this case we have $\hat{\pi}(\varepsilon) = \pi'\pi(\varepsilon)$ and if, for instance, the sample beam is unpolarized, π' acts like and unity operator, whence the formal analogy between the emission and absorption rate equations is complete. Essentially the same holds in cases, where π' equals $\pi(\varepsilon)$, being a projector onto the coherent polarization state ε, or where the detector is not sensitive to the polarization of the beam.

5. State multipoles and signal parameters

5.1. The generalized signal function

The rate equations (67, 68) constitute the fundamental set of equations for signal detection in optical pumping of dilute gases using conventional light sources. In this paper our intention is to point out the dependence of the fluorescence and transmission signals

[8]) There is a conceptual difference between both the polarization operators π or π' of the pumping or of the sample light beam and the polarization operator $\pi_e(\varepsilon)$, which describes the polarization of the signal detector. π or π' is a statistical operator which, in any case, satisfies the normalization condition $\text{tr}\,(\pi) = 1$. $\langle\sigma|\ \pi_e(\varepsilon)\ |\sigma'\rangle$ can be seen as the polarization sensitivity matrix of the detector. If $\pi_e(\varepsilon)$ is the projector onto a certain coherent polarization state, the trace normalization for $\pi_e(\varepsilon)$ holds, too, but if the detector is not sensitive to the polarization of the beam as in the emission rate example stated below, we have $\text{tr}\,(\pi_e\ (\text{unpol.})) = 2$.

upon the atomic state multipoles under the various conditions of observation, in particular, the angular dependence and the influence of the polarization parameters. We can treat this problem under a common point of view, if we introduce a generalized signal function $\mathscr{S}$ according to:

$$\mathscr{S} = \mathrm{tr}\,(\varrho\pi).$$

Obviously, this is a simplified notation of the corresponding terms in (67) and (68). It should be pointed out here, that the generalized signal function does not contain that part of information, which is governed by the excitation probability factor P_b according to (52). But this contribution to the fluorescence signal does not depend upon the special conditions of observation (direction and polarization) and can be treated as an intensity factor, that is determined by the pumping cycle alone. The same holds, in principle, for the transmission signal and the ground state probability factor P_a. But since according to (53) $P_a \equiv 1$ is a good approximation, this modification here is of minor interest. The transition to the spherical tensor notation is straight-forward. Regarding (A9) we fit close to the foregoing representation if we give the tensor components of $\varrho(S)$ and $\varrho(A)$ in terms of the atomic state multipoles. From the definitions (60) and the relations (A10) we obtain:

$$\varrho_{K,\varkappa}(S) = C_K(b, a)\, \varrho_{K,\varkappa}(b)$$

with

$$C_K(b, a) = (-1)^{a+b+1+K}\, \hat{b}^2 \begin{Bmatrix} 1 & 1 & K \\ b & b & a \end{Bmatrix}$$

and

$$\varrho_{K,\varkappa}(A) = C_K(a, b)\, \varrho_{K,\varkappa}(a) \tag{69}$$

with

$$C_K(a, b) = (-1)^{a+b+1}\, \hat{a}^2 \begin{Bmatrix} 1 & 1 & K \\ a & a & b \end{Bmatrix}.$$

$\varrho(a)$ or $\varrho(b)$ is the statistical operator of the ensemble of atoms in the ground or excited state and the coefficients C_K are proportional to a 6-j-symbol, for which we used the notation of Ref. [*10*]. With (69) and (A9) the generalized signal function reads:

$$\mathscr{S} = \mathrm{tr}\,(\varrho\pi) = \sum_{K,\varkappa} C_K \varrho^*_{K,\varkappa} \pi_{K,\varkappa}. \tag{70}$$

$\varrho_{K,\varkappa}$ stands for the atomic state multipoles $\varrho_{K,\varkappa}(a)$ or $\varrho_{K,\varkappa}(b)$, respectively, and the $\pi_{K,\varkappa}$ are the spherical tensor operator components of the polarization operator $\pi(\varepsilon)$ or $\hat{\pi}(\varepsilon)$. With reference to similar problems in gamma-ray spectroscopy we call $C_K\pi_{K,\varkappa}$ the radiation parameters of optical pumping signals [*28*].
In the following sections we give an outline of the main properties of the radiation parameters, starting in Sec. 5.2 with a discussion of the general structure of the polarization matrix π_e. The result will be a reduction of the general scope to some basic situations, where it is sufficient to consider coherent states of polarization, only. For these situations we specify the radiation parameters in Sec. 5.3 and make a general comment on the detection of atomic state multipoles. In Sec. 5.4 we obtain a survey of the angular dependence of the radiation parameters according to the transformation (34) and, finally write down explicit expressions in the Euler angles for the basic components of the generalized signal function.

5.2. Basic polarization states and the expansion of the polarization matrix

We refer to Sec. 3.1 where we introduced a basic set of circularly polarized photon states. For the one-quantum states (15) with wave vector $\boldsymbol{k} = k\boldsymbol{e}$ we use the simplified notation:

$$|\sigma\rangle = a^+_{k,\sigma}\,|\text{vac}\rangle = |1, \boldsymbol{k}, \sigma\rangle \qquad \sigma = \pm 1 \tag{71}$$

and the polarization operator π_e according to (23) can be treated as an operator acting in a state space of total angular momentum quantum number $j = 1$.
Hence, for the basic polarization states σ, the matrix representation of the polarization operators $\pi_e(\sigma)$ are:

$$\begin{aligned}\langle\lambda|\,\pi_e(+1)\,|\lambda'\rangle &= \langle\lambda|\,a^+_{+1}a_{+1}\,|\lambda'\rangle\colon \begin{pmatrix}1&0&0\\0&0&0\\0&0&0\end{pmatrix}\\ \langle\lambda|\,\pi_e(-1)\,|\lambda'\rangle &= \langle\lambda|\,a^+_{-1}a_{-1}\,|\lambda'\rangle\colon \begin{pmatrix}0&0&0\\0&0&0\\0&0&1\end{pmatrix}.\end{aligned} \tag{72}$$

In these matrices, the successive lines correspond to $\lambda = +1, 0, -1$ and the columns are arranged in the same manner from left to right.
For a linearly polarized field, whose plane of polarization is determined by the direction of propagation $\boldsymbol{e}$ and an unit vector $\boldsymbol{e}_{\varphi_0}$ (cf. (20) and Fig. 1)

$$\boldsymbol{e}_{\varphi_0} = \boldsymbol{e}_{x'} \cos \varphi_0 + \boldsymbol{e}_{y'} \sin \varphi_0,$$

the one-quantum state (21) reads:

$$\begin{aligned}|\varphi_0\rangle &= (2)^{-1/2}\,(-\exp(-i\varphi_0)\,|+1\rangle \\ &\qquad + \exp(i\varphi)\,|-1\rangle) \\ &= \sum_\sigma |\sigma\rangle\,\langle\sigma \mid \varphi_0\rangle.\end{aligned} \tag{73}$$

For the polarization operator $\pi_e(\varphi_0)$:

$$\pi_e(\varphi_0) = a^+_{\varphi_0} a_{\varphi_0} \sum_{\sigma,\sigma'} a_\sigma^+ \langle\sigma \mid \varphi_0\rangle\,\langle\varphi_0 \mid \sigma'\rangle\,a_{\sigma'} \tag{74}$$

we obtain the matrix representation:

$$\langle\lambda|\,\pi_e(\varphi_0)\,|\lambda'\rangle\colon (1/2)\begin{pmatrix}1 & 0 & -\exp(-2i\varphi_0)\\ 0 & 0 & 0\\ -\exp(2i\varphi_0) & 0 & 1\end{pmatrix}. \tag{75}$$

With the results of this summary let us now look at the general structure of the polarization matrices π_e, used in this context.
The projectors $\pi_e(\varepsilon)$ ($\varepsilon = \sigma, \varphi_0$) and the statistical operator π_e' of the sample beam are hermitian and because of the transversality of the field, all matrix elements $\langle\lambda|\,\pi_e\,|\lambda'\rangle$ vanish, for which either λ or λ' equals O. This latter property is conserved in summation or multiplication of those matrices and thus holds for the hermitian operator $\hat{\pi}_e(\varepsilon)$ in (68), too. Hence the general form of the polarization matrix π_e is:

$$\pi_e\colon \begin{pmatrix} p_1 & 0 & p_3 \exp(-2i\varphi)\\ 0 & 0 & 0\\ p_3 \exp(2i\varphi) & 0 & p_2\end{pmatrix}, \tag{76}$$

$p_\nu (\nu = 1, 2, 3)$ being a set of three real numbers.

If we identify π_e with $\pi_e(\varepsilon)$, we have to bear in mind that, in general, the parameters p_ν of this representation depend on the factor $T_p' \cdot \Delta E_p'$ in the definition (68). The same holds for the "effective polarization angle" φ. Thus, in this case, the numbers p_1 and p_2 are not coupled by a normalization condition nor can the angle φ simply be identified with the position angle of a polarizer actually used in the experimental setup.
However, the representation (76) allows a general treatment by means of some standard cases noting that, obviously, the matrices (72, 75) constitute a complete basis for an expansion of the polarization matrix:

$$\pi_e = c_{+1}\pi_e(+1) + c_{-1}\pi_e(-1) + c_\varphi \pi_e(\varphi) \tag{77}$$

with

$$c_{+1} = p_1 + p_3; \quad c_{-1} = p_2 + p_3; \quad c_\varphi = -2p_3 .$$

Using this expansion, the generalized signal function (70) can be written as a sum over basis functions $\mathcal{S}(\varepsilon)$ ($\varepsilon = \sigma, \varphi$):

$$\mathcal{S} = c_{+1}\mathcal{S}(1) + c_{-1}\mathcal{S}(-1) + c_\varphi \mathcal{S}(\varphi) \tag{78}$$

with

$$\mathcal{S}(\varepsilon) = \operatorname{tr}\left(\varrho\pi(\varepsilon)\right) \qquad \varepsilon = \sigma, \varphi ,$$

and we can restrict our further considerations to the basic polarisation matrices $\pi_e(\varepsilon)$ according to (72) and (75).

5.3. Polarization state multipoles

From the transformation formulae between the matrix elements and the spherical tensor operator components, in particular, for the polarization operator we obtain (cf. (A10)):

$$\tilde{\pi}_{K,\varkappa}(\varepsilon) = \sum_{\lambda,\lambda'} \langle\lambda|\, \tilde{\pi}(\varepsilon)\, |\lambda'\rangle\, (-1)^{1-\lambda} \cdot \langle 1, \lambda'; 1, -\lambda \mid K, \varkappa\rangle . \tag{79}$$

Since throughout in this section we relate representations to the signal frame, we left the index e but use a tilde in order to distinguish this representation from that in the atomic frame.
The polarization operator $\tilde{\pi}(\varepsilon)$ ($\varepsilon = \sigma, \varphi$) is a projector onto photon states of certain polarization and with reference to the designation used in connection with atomic states, we call $\tilde{\pi}_{K,\varkappa}(\varepsilon)$ the polarization state multipoles.
For the basic cases (72, 75) the polarization state multipoles are:

circular:

$$\begin{aligned} \tilde{\pi}_{1,\varkappa}(\sigma) &= \sigma(2)^{-1/2}\,\delta_{\varkappa,0} \\ \tilde{\pi}_{2,\varkappa}(\sigma) &= (6)^{-1/2}\,\delta_{\varkappa,0} \\ \sigma &= \pm 1 \end{aligned} \tag{80}$$

linear:

$$\begin{aligned} \tilde{\pi}_{1,\varkappa}(\varphi) &= 0 \quad \text{for all } \varkappa \\ \tilde{\pi}_{2,0}(\varphi) &= (6)^{-1/2} \\ \tilde{\pi}_{2,\pm 1}(\varphi) &= 0 \\ \tilde{\pi}_{2,\pm 2}(\varphi) &= -(2)^{-1} \exp(\pm 2i\varphi) . \end{aligned}$$

Independent of the special representation we have the normalization condition:

$$\tilde{\pi}_{0,0}(\varepsilon) = 3^{-1/2} \qquad (\varepsilon = \sigma, \varphi) \tag{81}$$

and the conjugation relation:

$$\tilde{\pi}_{K,-\varkappa}(\varepsilon) = (-1)^{\varkappa}\, \tilde{\pi}^{*}_{K,\varkappa}(\varepsilon). \tag{82}$$

With these results, we are now in a position to make some general comments on the observation of atomic orientation parameters, which are the state multipoles $\varrho_{K,\varkappa}(j)$ $(j = a, b)$. Using the fact that the transformation rules for the irreducible tensor operator components does not mix different tensor order K, these considerations can illustrate the advantages of the spherical tensor notation.

First of all, the optical detection signals carry direct informations about such atomic state multipoles only, for which $K \leqq 2$. Higher order tensor orientation parameters can affect the optical detection signal only as much as their dynamics act on the dynamics of the state multipoles with tensor order $K \leqq 2$ according to the kinetic equations of optical pumping (44, 45).

Further, from (80) and the structure of the generalized signal function (70) one will conclude, that a linearly polarized fluorescence signal will contain no information about the excited state orientation parameters $\varrho_{1,\varkappa}(b)$; in order to observe these state multipoles one has to use a circularly polarized detection channel [7].

However, independent of the polarization used in the signal channel, there is an over-all contribution of the ground state parameters to the fluorescence signal that accounts for the ground state multipole dependence of the excitation probability P_b (cf. (53)). But since all considerations concerning the polarization state multipoles can be applied to the main pumping beam, one has to use a circularly polarized pumping beam in order to produce an orientation $\varrho_{1,\varkappa}(a)$ and thus observe a contribution of these parameters to the fluorescence signal.

This simple relationship between tensor order K of the observed atomic orientation parameters and the polarization of the radiation field, in general, does not hold for the transmission signal. Because of the more complicated structure of the operator $\tilde{\pi}(\varepsilon)$ in (68) there might be, for instance, a signal contribution of the orientation parameters with $K = 1$ due to virtual transitions in cases, where both, the sample beam and the detection channel, are linearly polarized but with different planes of polarization.

Those general criteria for the observation of atomic orientation parameters must be completed by consideration of the angular dependence of the signal function. In our context, this angular dependence is determined by the transformation of the polarization state multipoles, or else the radiation parameters, from the signal to the atomic frame. In the following section we derive explicit expressions in the Euler angles for the radiation parameters and give a graphical survey of these features.

5.4. The angular dependence of the generalized signal function

In order to describe the polarization of radiation in the direction of observation $\boldsymbol{e}$, we choose a cartesian triad $\boldsymbol{e}_{x'}$, $\boldsymbol{e}_{y'}$, $\boldsymbol{e}_{z'}$ with $\boldsymbol{e}_{z'} = \boldsymbol{e}$. This triad, which we call the signal frame, may be obtained from the original frame $\boldsymbol{e}_x$, $\boldsymbol{e}_y$, $\boldsymbol{e}_z$, related to the atomic states, by a rotation $\mathcal{R}(\alpha, \beta, \gamma)$, wherre α, β, γ are the corresponding Euler angles.

In the rotation $\mathcal{R}(\alpha, \beta, \gamma)$, which rotates the atomic triad into the signal frame (cf. Fig. 1), the polarization state multipoles transform according to:

$$\pi_{K,\varkappa}(\varepsilon) = \sum_{\varkappa'} \tilde{\pi}_{K,\varkappa'}(\varepsilon)\, R^{(K)*}_{\varkappa,\varkappa'}(\alpha, \beta, \gamma), \tag{83}$$

where $\pi_{K,\varkappa}(\varepsilon)$ are the polarization state multipoles in the atomic frame and

$$R^{(K)*}_{\varkappa,\varkappa'}(\alpha, \beta, \gamma) = \exp(i\varkappa\alpha)\, r^{(K)}_{\varkappa,\varkappa'}(\beta)\, \exp(i\varkappa'\gamma) \tag{84}$$

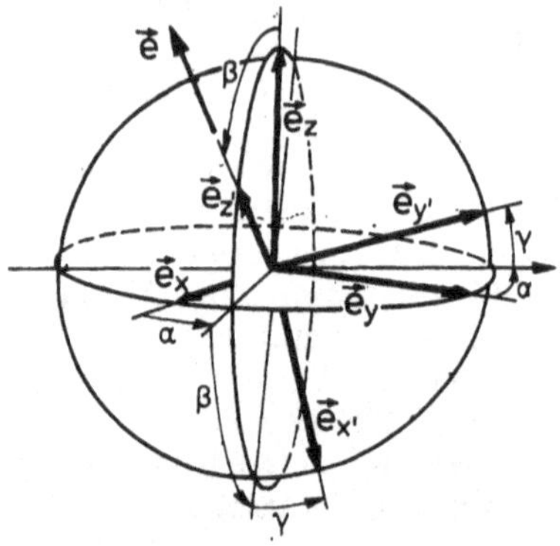

Fig. 1. Rotation of the atomic frame into the signal frame, specified by the Euler angles α, β, γ

is a $(2K + 1)$-dimensional standard representation of this rotation[9]). Inserting for the $\bar{\pi}_{K,\varkappa}(\varepsilon)$ according to (80), we obtain, in particular, for circular polarization:

$$\begin{aligned}
\pi_{1,\varkappa}(\pm 1) &= \pm(2)^{-1/2}\, R^{(1)*}_{\varkappa,0}(\alpha, \beta, \gamma)\\
\pi_{1,1}(\pm 1) &= \mp(1/2)\exp(i\alpha)\sin\beta\\
\pi_{1,0}(\pm 1) &= \pm(2)^{-1/2}\cos\beta\\
\pi_{2,\varkappa}(\pm 1) &= (6)^{-1/2}\, R^{(2)*}_{\varkappa,0}(\alpha, \beta, \gamma)\\
\pi_{2,2}(\pm 1) &= (1/4)\exp(i\,2\alpha)\sin^2\beta\\
\pi_{2,1}(\pm 1) &= -(1/2)\exp(i\alpha)\sin\beta\cos\beta\\
\pi_{2,0}(\pm 1) &= -(1/2)\,(6)^{-1/2}\,(1 - 3\cos^2\beta)
\end{aligned} \tag{85}$$

and for linear polarization we find:

$$\pi_{1,\varkappa}(\varphi) = 0 \qquad \text{for all } \varkappa$$

$$\pi_{2,\varkappa}(\varphi) = \exp(i\varkappa\alpha) \times \Big[(6)^{-1/2}\, r^{(2)}_{\varkappa,0}(\beta) - (1/2)\Big(\exp\big(i\,2(\gamma+\varphi)\big)\, r^{(2)}_{\varkappa,2}(\beta) + \exp\big(-i\,2(\gamma+\varphi)\big)\, r^{(2)}_{\varkappa,-2}(\beta)\Big)\Big]. \tag{86}$$

As would be expected from the very definition, the polarization angle φ in (86) occurs like an additional rotation of the signal frame around the direction of observation $\boldsymbol{e}$. Since such a rotation is optional, we can, without loss of generality, restrict considerations to the case, where $\varphi = 0$:

$$\begin{aligned}
\pi_{2,2}(0) &= (1/4)\exp(i\,2\alpha)\big(\sin^2\beta - (1 + \cos^2\beta)\cos 2\gamma - i\,2\cos\beta\sin 2\gamma\big)\\
&= (1/2)\exp(i\,2\alpha)\big(1 - (1 + \cos^2\beta)\cos^2\gamma - i\,2\cos\beta\sin\gamma\cos\gamma\big)
\end{aligned} \tag{87}$$

[9]) As for the rotation matrix $r^{(K)}_{\varkappa,\varkappa'}(\beta)$ confer, for instance to [*10*, *29*].

$$\pi_{2,1}(0) = -(1/2)\exp(i\alpha)\big(\sin\beta\cos\beta\,(1+\cos 2\gamma) + i\sin\beta\sin 2\gamma\big)$$
$$= -\exp(i\alpha)\,(\sin\beta\cos\beta\cos^2\gamma + i\sin\beta\sin\gamma\cos\gamma)$$
$$\pi_{2,0}(0) = (6)^{-1/2}\,(1 - 3\sin^2\beta\cos^2\gamma). \tag{87}$$

As for the fluorescence signal, the restriction to $\varphi = 0$ implies a quite natural choice of the signal frame, so that the x', z'-plane coincide with the plane of polarization of the signal channel. Regarding the transmission signal, there are several cases, for which this restriction seems to be quite natural, too, but it could seem to be more sophisticated in cases, where the virtual transition term contributes to π_e and hence determines the effective polarization angle φ according to the decomposition (77). However, since there is a simple return to the general case, we may adopt, for convenience, the condition $\varphi = 0$ in our further representation.
Furthermore, the Euler angle α occurs in the phase factor $\exp(i\varkappa\alpha)$, only, which describes a rotation in the complex $\pi_{K,\varkappa}$-plane, and does not affect the order of magnitude of the signal parameters. For the graphical representation we therefore use the decomposition:

$$\pi_{K,\varkappa}(\varepsilon) = \exp(i\varkappa\alpha)\, s_{K,\varkappa}(\varepsilon) \quad \text{with} \quad s_{K,-\varkappa}(\varepsilon) = (-1)^{\varkappa}\, s^*_{K,\varkappa}(\varepsilon). \tag{88}$$

In case of circular polarization, the parameters $s_{K,\varkappa}(\pm 1)$ are real functions in the Euler angle β and we have:

$$s_{K,\varkappa}(-1) = (-1)^K\, s_{K,\varkappa}(+1). \tag{89}$$

The angular dependence of $s_{K,\varkappa}(+1)$ is shown in Fig. 2. From these diagrams, for instance one easily derives conditions of observation, which allow to discriminate between the signal contribution of orientation parameters $\varrho_{K,\varkappa}(j)$ with different values for $\varkappa$.
In case of linear polarization, the parameters $s_{2,\varkappa}(0)$ with $\varkappa \neq 0$, in general, are complex. They can be represented in the form:

$$s_{2,\varkappa}(0) = \hat{x}_\varkappa(\beta) - \big(x_\varkappa(\beta,\gamma) + i y_\varkappa(\beta,\gamma)\big) \tag{90}$$

with

$$\frac{x_\varkappa{}^2}{A_\varkappa{}^2(\beta)} + \frac{y_\varkappa{}^2}{B_\varkappa{}^2(\beta)} = 1 \qquad \varkappa = 1, 2.$$

For each value of β, the function $s_{2,\varkappa}(0)$ in the Euler angle γ represents an ellipse in the complex plane with its centre on the real axis at $\hat{x}_\varkappa(\beta)$ and with the semi-axis $A_\varkappa(\beta)$ and $B_\varkappa(\beta)$. The graphs of the parameters $s_{2,\varkappa}(0)$ are shown in Fig. 3. They can be used, for instance, to check the order of magnitude in the contribution to the signal function $\mathcal{S}(0)$.
We want to complete our representation with explicit expressions for the basic signal functions $\mathcal{S}(\varepsilon)$ ($\varepsilon = \sigma, \varphi$), introduced with (78), and again adopt the condition $\varphi = 0$. Since the signal function $\mathcal{S}$ is real, we pass over to a real notation using the decomposition:

$$\varrho_{K,\varkappa} = u_{K,\varkappa} + i v_{K,\varkappa} \tag{91}$$

for the atomic state multipoles. Inserting for the polarization state multipoles we obtain from (78) together with (70, 85, 87):

$$\begin{aligned}\mathcal{S}(\pm 1) = {} & C_0 3^{-1/2} u_{0,0}\\ & + C_1\big(\pm(2)^{-1/2}\cos\beta u_{1,0} \mp (\cos\alpha\sin\beta u_{1,1} + \sin\alpha\sin\beta v_{1,1})\big)\\ & + C_2\big(-(1/2)\,6^{-1/2}(1 - 3\cos^2\beta)\, u_{2,0}\\ & - (1/2)\,(\cos\alpha\sin 2\beta u_{2,1} + \sin\alpha\sin 2\beta v_{2,1})\\ & + (1/2)\,(\cos 2\alpha\sin^2\beta u_{2,2}\sin 2\alpha\sin^2\beta v_{2,2})\big)\end{aligned} \tag{92}$$

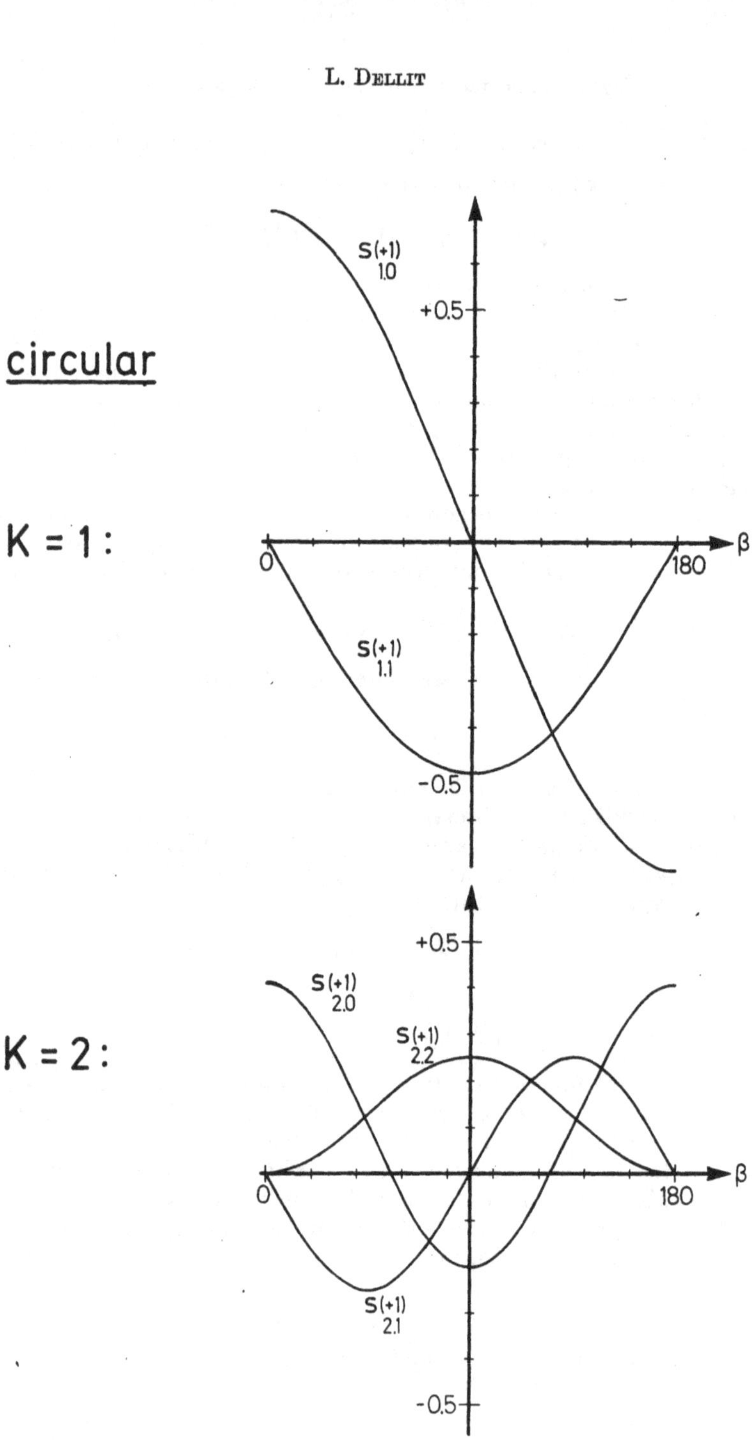

Fig. 2. Angular dependence of the polarization parameters $s_{K,\varkappa}(+1)$

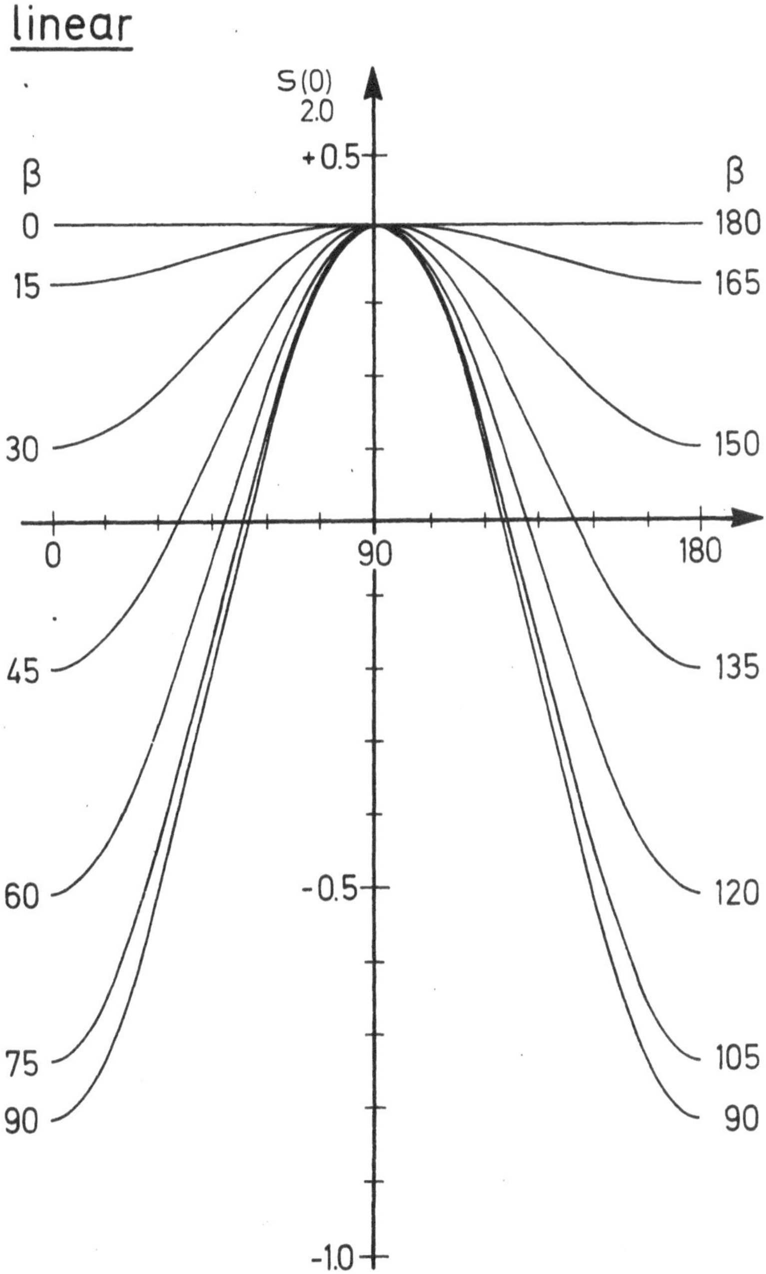

Fig. 3a. Angular dependence of the polarization parameter $s_{2,0}(0)$

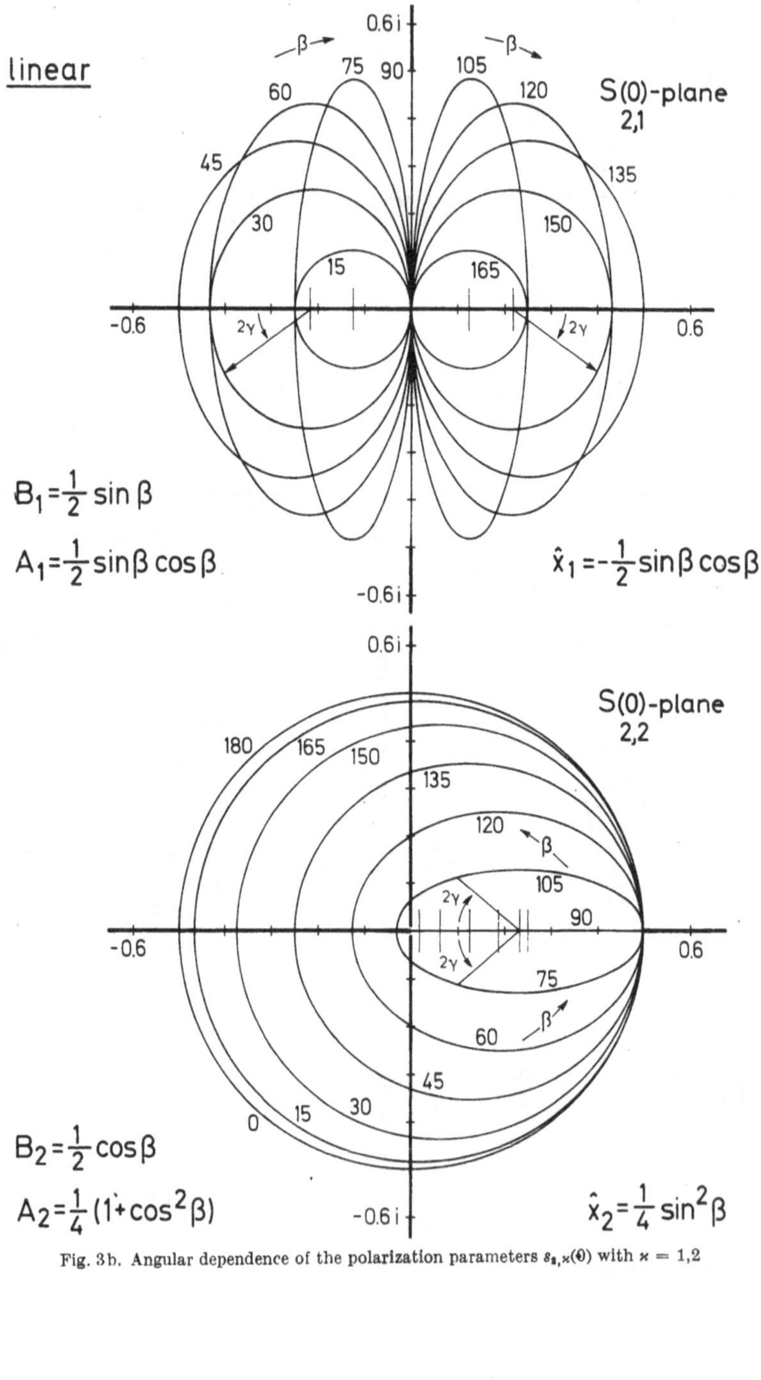

Fig. 3b. Angular dependence of the polarization parameters $s_{2,\varkappa}(\Theta)$ with $\varkappa = 1,2$

and

$$\begin{aligned}\mathcal{S}(0) = C_0 3^{-1/2} u_{0,0} + C_2 \Big(6^{-1/2}(1 - 3\sin^2\beta\cos^2\gamma)\, u_{2,0} \\ - (\cos\alpha\sin 2\beta\cos^2\gamma - \sin\alpha\sin\beta\sin 2\gamma)\, u_{2,1} \\ - (\sin\alpha\sin 2\beta\cos^2\gamma + \cos\alpha\sin\beta\sin 2\gamma)\, v_{2,1} \\ + \big(\cos 2\alpha(1 - (1+\cos^2\beta)\cos^2\gamma) + \sin 2\alpha\cos\beta\sin 2\gamma\big)\, u_{2,2} \\ + \big(\sin 2\alpha(1 - (1+\cos^2\beta)\cos^2\gamma) - \cos 2\alpha\cos\beta\sin 2\gamma\big)\, v_{2,2}\Big). \qquad (93)\end{aligned}$$

These are the basic components of the generalized signal function according to (70, 78), that describes the optical detection signal in optical pumping experiments with dilute gases using a non coherent light source. The $u_{K,\varkappa}$ and $v_{K,\varkappa}$ stand for the real and imaginary part of the generally complex atomic state multipoles, which are the orientation parameters of the ground or excited state, respectively. The parameters C_K are, essentially, the vector coupling coefficients (6-j-symbols) of (69) expressing the conservation of total angular momentum in the elementary process of absorption and emission. The direction of observation with respect to the atomic frame is determined by the Euler angles α and β and in the case of linear polarization, the position of the plane of polarization is given by the Euler angle γ. In general, the signal function $\mathcal{S}$ is a real linear superposition of the functions (92, 93), but in many experimental situations one would come out with one of these basic functions, alone.

5. Conclusions

During the last years optical pumping has become an efficient method of investigation and made an important contribution to the renascence in atomic physics. In many cases it set up a new approach to elder problems.

Particular, in the context of atomic collision processes and in the description of relaxation phenomena the irreducible tensor notation, a standard tool in nuclear physics, came more and more into use [*31*]. This development seems to be quite natural under the uniform aspect of radiation interacting whit matter and the classification of interaction with respect to spatial symmetries. So it seemed to be useful to give a formulation of the cycle of optical pumping itself in irreducible tensor notation.

The long-wavelength approximation and the limitation to electric dipole transitions simplifies the treatment of radiation parameters and allows the derivation of an efficient concept of polarization by means of the statistical operator of the pumping light beam generated by a lamp with thermal-like statistics.

The introduction of irreducible tensor notation in our context is a posteriori. We used this approach in order to point out some aspects in connection with beam divergencies and a possible distortion of symmetries in the extended magnetic field case. More general, one could start with an irreducible tensor expansion of the Liouville equation itself (cf. [*19*]) using all the a priori symmetries of the problem.

The kinetic equations for the irreducible tensor components of the statistical operator of the atomic system give a comprehensive description of the optical pumping process. They constitute a set of interdependent differential equations, but since the formulation is fit to spatial symmetries, the decoupling is carried out as far as is possible in the stated experimental situation.

Moreover, the treatment of optical signal parameters in irreducible tensor notation deduced from the rate equations for spontaneous emission and absorption illustrate the advantages of this formalism. Since the signal function, which carries information about the state multipoles and hence informations about the dynamics of atomic orientation, is mainly determined by the geometrical conditions of observation, the spherical tensor operator formalism proves to be an efficient tool in handling with those problems. This geometrical aspect led to a unified treatment of fluorescence and transmission signals by means of a generalized signal function, with minor emphasis on differences, which otherwise would probably be of physical interest.

Acknowledgement

I want to express my gratitude to Prof. U. Cappeller for his stimulating interest in this work and for many discussions. I am also indebted to Prof. W. Walcher, who gave me the earlier opportunity of working in this field.

Appendix: The multipole operators $U_\varkappa^{(K)}(j, j')$

By definition, an irreducible spherical tensor operator $T^{(K)}$ of order K is a set of $2K + 1$ operators $T_\varkappa^{(K)}(\varkappa = -K, \ldots, K)$ that transform under rotation according to:

$$RT_\varkappa^{(K)}R^{-1} = \sum_{\varkappa'} T_{\varkappa'}^{(K)} R_{\varkappa',\varkappa}^{(K)}, \tag{A1}$$

where $R_{\varkappa',\varkappa}^{(K)} = \langle K, \varkappa'| R |K, \varkappa\rangle$ is a standard representation of the rotation R (rotation matrix) specified by the corresponding Euler angles [*10*]. The algebra of these operators is determined by the commutation relations with the operator J of infinitesimal rotation

$$\begin{aligned} [J_\pm, T_\varkappa^{(K)}] &= \big(K(K+1) - \varkappa(\varkappa \pm 1)\big)^{1/2} T_{\varkappa\pm1}^{(K)} \\ [J_z, T_\varkappa^{(K)}] &= \varkappa T_\varkappa^{(K)}. \end{aligned} \tag{A2}$$

A special kind of those operators was introduced by Fano [*19, 30*] who called them multipole operators. These multipole operators $U_\varkappa^{(K)}(j, j')$ are given by the matrices:

$$\langle j'', m| U_\varkappa^{(K)}(j, j') |j''', m'\rangle = \delta_{j'',j}\delta_{j''',j'}(-1)^{j'-m'} \langle j, m; j', -m' |K, \varkappa\rangle \tag{A3}$$

where K runs from $|j - j'|$ to $j + j'$. A comparison with the Wigner-Eckart theorem (30) shows that $U^{(K)}(j, j')$ is an irreducible tensor operator whose reduced matrix element is the mere normalization factor $\hat{K}$.

Beside the fundamental equations (A1) and (A2) we have the trace relation:

$$\operatorname{tr}\big(U_\varkappa^{(K)}(j, j')\big) = \delta_{j,j'}\delta_{K,0}\delta_{\varkappa,0}\hat{j} \tag{A4}$$

and the $U_\varkappa^{(K)}$ obey the 'orthogonality' condition

$$\operatorname{tr}\big(U_\varkappa^{(K)+}(j, j')\, U_\lambda^{(L)}(j'', j''')\big) = \delta_{j,j''}\delta_{j',j'''}\delta_{K,L}\delta_{\varkappa,\lambda} \tag{A5}$$

with the conjugation relation:

$$U_\varkappa^{(K)+}(j, j') = (-1)^{j'-j}\,(-1)^\varkappa\, U_{-\varkappa}^{(K)}(j', j). \tag{A6}$$

Further, in connection with the spherical tensor expansion of the operator $\mathscr{A}(\lambda, \lambda') = \hat{1}^{-2}U_{\lambda'}^{(1)+}(b, a)\, U_\lambda^{(1)}(b, a)$ there is the useful product relation:

$$U_\varkappa^{(K)}(j, j')\, U_\lambda^{(L)}(j', j) = \sum_{M,\mu} (-1)^{2j}\,\hat{K}\hat{L}(-1)^M \langle K, \varkappa; L, \lambda \,|\, M, \mu\rangle \begin{Bmatrix} M & K & L \\ j' & j & j \end{Bmatrix} U_\mu^{(M)}(j, j). \tag{A7}$$

The set of multipole operators is 'complete' in the sense, that any matrix $\langle j, m| Q |j', m'\rangle$ can be expanded into a sum of matrices (A3). In particular, the density matrix μ_{at} can be represented by such an expansion and according to (38) we are mainly interested in the case $j = j'$. For simplicity, we write $U_\varkappa^{(K)}(j)$ instead of $U_\varkappa^{(K)}(j, j)$ and obtain for the statistical operator of the atomic ground or excited, state, respectively:

$$\varrho(j) = \sum_{K,\varkappa} \varrho_{K,\varkappa}(j)\, U_\varkappa^{(K)+}(j) \tag{A8}$$

with

$$\varrho_{K,\varkappa}(j) = \mathrm{tr}\left(\varrho(j)\, U_\varkappa^{(K)}(j)\right) \qquad j = a, b.$$

The coefficients $\varrho_{K,\varkappa}(j)$ of this expansion are the mean values of the multipole operators carried out on the atomic system in the ground or excited state. Since $\varrho(j)$ is a statistical operator in the state space $|j, m\rangle$, the $\varrho_{K,\varkappa}(j)$ are sometimes called state multipoles or else irreducible tensor operator components.

The state multipoles constitute another tensorial set which describes the system in the state space $|j, m\rangle$ as completely as the matrix elements $\langle j, m| \varrho(j) |j, m'\rangle$ of the statistical operator. Consider any other observable $Q(j)$ of the system expanded according to (A8). Then for the expectation value $\langle Q(j)\rangle$ we have:

$$\langle Q(j)\rangle = \mathrm{tr}\left(Q(j)\, \varrho(j)\right) = \sum_{K,\varkappa} q^+_{K,\varkappa}(j)\, \varrho_{K\varkappa,}(j) \tag{A9}$$

with

$$q^+_{K,\varkappa}(j) = \mathrm{tr}\left(Q(j)\, U_\varkappa^{(K)+}(j)\right) = (-1)^\varkappa\, q_{K,-\varkappa}(j)$$

and the hermiticity of $Q(j)$ implies: $q^+_{K,\varkappa}(j) = q^*_{K,\varkappa}(j)$ (complex conjugate).

Finally, the transformation formulae between the matrix elements and the irreducible tensor operator components of an operator $\mu(j)$ follow directly from the definitions and the orthogonality relations of the Clebsch-Gordon coefficients:

$$\begin{aligned} \mu_{K,\varkappa}(j) &= \sum_{m,m'} \langle j, m| \mu(j) |j, m'\rangle\, (-1)^{j-m} \langle j, m'; j, -m \mid K, \varkappa\rangle \\ \langle j, m| \mu(j) |j, m'\rangle &= \sum_{K,\varkappa} (-1)^{j-m} \langle j, m'; j, -m \mid K, \varkappa\rangle\, \mu_{K,\varkappa}(j). \end{aligned} \tag{A10}$$

References

[1] Barrat, J. P., Cohen-Tannoudji, C., J. Phys. Rad. **22**, 329, 443 (1961).
[2] Cohen-Tannoudji, C., Ann. Phys. (Paris) **7**, 423, 469 (1962).
[3] Happer, W., Rev. Mod. Phys. **44**, 169 (1972).
[4] Cappeller, U., Dellit, L., Phys. Letters **26 A**, 535 (1968).
[5] Cappeller, U., Dellit, L., Proceedings of the International Conference on Optical Pumping and Atomic Line Shape (O PaLS), 393, Warsaw (1968).
[6] Willis, Ch. R., Phys. Rev. **A1**, 467 (1970).
[7] D'yakonov, M. I., Soviet. Physics-JETO **20**, 1484 (1965).
[8] Barrat, J. P., J. Phys. Rad. **20**, 541, 633, 657 (1959).
[9] Cohen-Tannoudji, C., Laloe, J., J. Phys. **28**, 505, 722 (1967).
[10] Messiah, A., Quantum Mechanics, Amsterdam, North-Holland Publishing Company 1965.
[11] Glauber, R. J., Phys. Rev. **131**, 2766 (1963).
[12] Mandel, L., Wolf, E., Rev. Mod. Phys. **37**, 231 (1965).
[13] Loudon, R., The Quantum Theory of Light, Oxford, Clarendon Press 1973.
[14] Shore, B. W., Menzel, D. H., Principles of Atomic Spectra, New York, John Wiley & Sons 1968.
[15] Edmonds, A. R., Angular Momentum in Quantum Mechanics, Princeton University Press.
[16] Happer, W., Mathur, B. S., Phys. Rev. **163**, 12 (1967).

[17] Dupont-Roc, J., Ledourneuf, M., Cohen-Tannoudji, C., II. Conference on Atomic Spectroscopy, Hannover (1970).
[18] Cappeller, U., Fortschr. Phys. **20**, 423 (1972).
[19] Fano, U., Rev. Mod. Phys. **29**, 74 (1957).
[20] Cagnac, B., Ann. Phys. (Paris) **6**, 467 (1961).
[21] Cohen-Tannoudji, C., J. Phys. **24**, 653 (1963).
[22] Haroche, S., Thèse de 3ième cycle, Paris (1967).
[23] Series, G. W., II Conference on Atomic Spectroscopy, Hannover (1970).
[24] Dellit, L., Dissertation, Marburg (1969).
[25] Cappeller, U., Weber, H. G., Ann. Phys. (Leipzig) **26**, 359 (1971).
[26] Shen, Y. R., Rev. Mod. Phys. **48**, 1 (1976).
[27] Weber, H.-G., Z. Phys. **247**, 336 (1971); with further references quoted there.
[28] Ferguson, A. J., Angular Correlation Methods in Gamma-Ray Spectroscopy; Amsterdam, North-Holland Publ. Comp. 1965.
[29] Brink, D. M., Satchler, G. R., Angular Momentum; London, Oxford University Press 1962.
[30] Fano, U., Phys. Rev. **90**, 577 (1953).
[31] Omont, A., Prog. Quantum Electronics, Vol. 5, pp. 69—138, Pergamon Press, 1977.

Fortschritte der Physik, 27, 35–74 (1979)

Consequences of Crossed Channel Structure for High Energy Elastic Scattering

OSWALD HAAN

Fachbereich 8 — Naturwissenschaften I, Gesamthochschule Wuppertal, BRD

Abstract

This review describes a field in the S-matrix theory of elastic stattering. Analyticity and elastic unitarity in the crossed t-channel are shown in a rigorous manner to yield constraints on s-channel quantities. Strong restrictions on s-channel partial wave with angular momenta $l \geqq \sqrt{s}/2(b_0 + \lambda \ln s)$ follow from realistic assumptions on the amplitude in the unphysical domain (Mandelstam analyticity and Regge behaviour). The possibility of phenomenological application of these results is discussed.

Contents

I. Introduction

Any description of strong interactions is based on the principles of causality, relativity and quantum mechanics. These fundamental properties of nature imply the following structure for the S-matrix [1], whose matrix elements are the amplitudes for scattering and production of hadrons:

1) S is invariant under Poincaré transformations,
2) S is unitary,
3) the amplitudes are boundary values of analytic functions,
4) one single analytic function describes, for different values of its arguments, both the interaction of particles and of their antiparticles, this is the crossing property.

It is the program of S-matrix theory [2] to formulate these properties in a precise way and to exploit their consequences for experimentally measurable quantities, without the use of further assumptions on the dynamics of strong interactions. The farthestreaching aim is the explicit construction of the amplitudes, with only few measurable masses and coupling-constances as input. However, the complexity of multiparticle amplitudes [*3*], which are all coupled in the unitarity equations, is a high barrier in the way of complete determination of strong interactions from basic principles.

Fortunately, unitarity also has consequences, which concern one single amplitude. The most important examples are the inequalities for the elastic amplitude. These unitarity inequalities, combined with the linear analyticity and crossing properties, give a very constraining structure for the elastic amplitude alone. Clearly, the amplitude will not be fixed from this partial use of basic principles, but it is possible to bound its magnitude, and to derive relations, involving the amplitude at different values of its variables.

The first result in this framework has been obtained by FROISSART in 1961 [*4*], an upper bound for the total cross-section of two particle scattering at high energy $\sqrt{s}$:

$$\sigma_{tot}(s) < C\left(\log\frac{s}{s_0}\right)^2, \tag{I.1}$$

where C and s_0 are undetermined constants.

Since then, many consequences of unitarity, analyticity and crossing have been derived, mainly by Martin and his collaborators [*5*]. Among the results are the determination of the constants C and s_0 in the Froissart-bound (I.1) and bounds on other quantities at high energies [*6*], as well as statements concerning the amplitude at low and medium energies [*7*]. There still is a great interest in new bounds and relations and in improvements of existing ones, because they find immediate applications in the phenomenological analysis of scattering experiments. Outstanding examples for these phenomenological applications are the scaling properties of diffraction scattering at high energies [*8*], the use of the Roy-equation [*9*] for the analysis of low energy $\pi-\pi$ scattering [*10*], and the stabilizing property of bounds for analytic continuation of given data [*11*].

Most of the results mentioned above and described in the reviews [*5, 6, 7, 10, 11*] have as a common feature that the analyticity properties of the amplitude, which are used for their derivation, follow from the principles of an axiomatic field theory. In particular, for the proof of the high energy bounds it suffices that the amplitude is analytic, as a function of the momentum transfer t, in the Lehmann-Martin-ellipse [*12*]. No use is made of the crossing property, that for real values of t outside this ellipse the amplitude describes another physical scattering process and that consequently in this domain of the variables, too, unitarity puts restrictions on the amplitude. It is therefore an obvious question, whether this additional information can be used to obtain stronger results for the scattering amplitude.

Already in 1958 MANDELSTAM [*13*] gave the equations, which express elastic unitarity in all scattering channels simultaneously. In this treatment the amplitude is assumed to be analytic in the maximal domain, which is still consistent with unitarity and crossing. Since this formulation of combined unitarity and crossing contains inelastic contributions which are not positive definite, it has not been possible to extract from it directly bounds for the amplitude in its physical domain. Instead the Mandelstam equations, sometimes in approximate form, have been used for a dynamical calculation of the amplitude, with special Ansätze for the inelastic contributions [*14*].

This program, in its most consequent form, has been and still is pursued by ATKINSON and collaborators [*15*] and by KUPSCH [*16*]. They derive an iterative procedure, which allows to construct amplitudes obeying all restrictions of unitarity, analyticity and crossing.

With these results the reduced aim of S-matrix theory is in principle reached, namely to exploit fully all properties of the elastic amplitude, following from basic principles. The iterative nature of the construction principle leads, however, to numerically small

amplitudes. Moreover, only amplitudes with at most constant cross sections have been considered so far. The scattering experiments, on the other hand, require large amplitudes and, at high energies, growing cross-sections. Therefore, this full solution of the S-matrix problem has so far no phenomenological consequences.
The above outlined two different mainstreams in S-matrix theory for elastic scattering can be characterized in the following way:

i) Martin's use of axiomatic analyticity leads to restrictions on the physical amplitude.
ii) The Atkinson-Kupsch approach, based on Mangelstam analyticity, allows the explicit construction of unrealistic small model amplitudes.

Another method to exploit S-matrix principles, which combines some features of i) and ii) will be described in this work. Following the Mandelstam-Atkinson-Kupsch line, this new approach uses analyticity assumptions, which are not justified in axiomatic field theory [cf. section II]; however no attempt is made to construct amplitudes fulfilling all S-matrix requirements. Instead, in the spirit of the axiomatic approach i), only some aspects of S-matrix principles will be implemented in such a way, that it is possible to treat realistic amplitudes. Most of the ideas and results [*27, 32, 34, 41*] which will be described in the next sections, originate from the collaboration with K. H. Mütter. Some contributions to the scaling law [cf. section III] are due to Common [*28, 31*] and to Yndurain [*29*].
The characteristic feature of this new line in S-matrix theory lies in the restriction to partial waves with high angular momenta l

$$l \geqq L(s) = \frac{\sqrt{s}}{2} (b_0 + \lambda \ln s), \tag{I.2}$$

for which consequences of properties in the crossed t-channel are considered. The choice of b_0 and λ will depend on the assumed magnitude of the amplitude in the unphysical domain, where energy s and momentum transfer t are positive. Essentially, this restriction to l-values much above the assumed position of the Regge trajectories allows to treat realistic amplitudes which are large enough and describe growing total cross-sections.
This use of crossed channel properties should be compared to the methods in the framework of axiomatic analyticity. There only properties of the amplitude below threshold in the crossed channel, as the positions of the threshold, or the magnitude of scattering lengths, are used in order to obtain restrictions on the partial waves with high values (I.2) of l.
New properties of the absorptive parts of the s-channel partial waves $\operatorname{Im} f_l(s)$ follow, if unitarity in the crossed t-channel at threshold is considered. A scaling law holds, according to which in the limit of large energy $\sqrt{s}$ and large angular momentum l $\operatorname{Im} f_l(s)$ is, up to a known factor, only dependent on l and s in the combination

$$x = l \cdot s^{-3/2}. \tag{I.3}$$

Furthermore, the x-dependence is given explicitely as an integral over the absorptive part of the amplitude at t-channel threshold. The domain of validity of this scaling law depends on the specific assumptions for the amplitude in the unphysical domain. The scaling law and its domain of validity will be discussed in section III.
Restrictions for partial waves involving finite values of l and s are obtained by considering crossed channel properties in a finite interval above threshold. In section IV, an approximate crossing sumrule is derived, which relates a weighted integral over one s-channel partial wave $\operatorname{Im} f_l(s)$, where $l = L(s)$ (cf. I.2), to weighted integrals over t-channel partial waves $\operatorname{Im} f_J(t)$, where $J \geqq \alpha_0 + \lambda \sqrt{t}$. It is the crucial point of this relation,

that the t-channel information is supplied in form of partial waves with their simple unitarity properties, which are measurable for even values of J (in the case of $\pi\pi$-scattering).

For the derivation of the crossing sumrule a completeness property of the Legendre functions of the second kind, $Q_J(z)$, is needed, which is proved in appendix B. This property guarantees, that any function of s with asymptotic behaviour s^{-J-1} can be approximated arbitrarily good by a linear combination of Legendre functions $Q_{J'}(1 + 2s/(t - 4))$ with $J' \geqq J$. This is the origin of the approximative character of the sumrule: the more t-channel partial waves are taken into account, the more precise the integral over the s-channel partial wave is reproduced.

It turns out, that the high energy behaviour of the s-channel partial wave is restricted, via the crossing sumrule, by the t-channel partial wave with angular momentum $J = \alpha_0 + \lambda\sqrt{t}$ as close as possible to the leading Regge singularity, since for such values of α_0 and λ the high energy contribution in the s-integral is strongly weighted. That means, that a t-channel partial wave with non-integer value of angular momentum is involved, which cannot directly by taken from experiment. However, a more extensive use of crossed channel structure leads to an approximative dispersion relation, which, together with unitarity, fixes the t-channel partial wave with this unphysical angular momentum. The approximative dispersion relation for the partial wave $f_{\alpha_0+\lambda\sqrt{t}}(t)$ will be derived in section V. It has the inportant property, that the contribution from the left hand cut is given completely in terms of s-channel partial waves with small angular momenta $l \leqq \sqrt{s}/2\,(b_0 + \lambda \ln s)$ and of t-channel partial waves with physical angular momenta. Again, the error involved in the approximation can be made as small as is required by taking into account a large enough number of physical t-channel partial waves. The dispersion integral over the right hand cut contains a factor $e^{-b_0\sqrt{t'}}$, for large values of t', which suppresses the inelastic region, if b_0 is chosen appropriately. Neglecting some corrections which can be expressed in terms of physical t-channel partial waves, one can thus state: The dispersion relation, together with elastic unitarity, allows to calculate the t-channel partial wave with angular momentum J near the leading Regge singularity from the s-channel partial waves with samll angular momenta $l < L(s)$. The knowledge of this t-channel partial wave restricts in turn, with the help of the crossing sumrule, the s-channel partial waves with large angular momenta $l \geqq L(s)$.

In view of this short description of the results to be presented in detail below, the following basic characterization of this work can be given: Within the framework of S-matrix theory for elastic scattering, properties of the amplitude in the elastic domain of the crossed t-channel can efficiently be used to find quantitative constraints on s-channel partial waves with high angular momenta.

For a t-channel description of s-channel partial waves with smaller angular momenta it would be necessary to consider the inelastic crossed channel contribution, too. This has, up to now, not been possible in a model-independent way. However, within a specific model, the reggeon field theory [17], this line of procedure has been pursued successfully.

In present day S-matrix theory, it is appropriate, to use for s-channel partial waves with small angular momenta the restriction of direct channel unitarity:

$$0 \leqq \operatorname{Im} f_l(s) \leqq \frac{\sqrt{s}}{s-4}. \tag{I.4}$$

This mixed view of partial waves, from the direct channel for small angular momenta, from the crossed channel for high angular momenta, has some support from phenomenology.

The upper s-channel bound (I.4) is nearly saturated for small l. However, for large l, the partial waves decrease rapidly, a behaviour, which can be described by use of t-channel unitarity, e.g. in the form of scaling law (III.4, III.6).
In the above discussion on the significance of crossing sumrule and dispersion relation the corrections due to their approximative nature had been neglected. It is true that these corrections can be expressed in terms of measurable t-channel partial waves, and that the relations can be made arbitrarily exact by taking into account enough of them. But in a phenomenological application of the relations only a fixed limited number of physical t-channel partial waves will be available and an estimate of the remaining error becomes important. This remaining error depends on the magnitude of the amplitude in the unphysical domain (cf. section IV, V). It has not been possible to derive bounds on the errors from S-matrix principles alone. Additional information on the amplitude in the unphysical domain is necessary for this. In section VI such assumptions, which concern the oscillations of the amplitude, and the resulting error bounds are discussed. It remains the important problem to check the overall consistency of those additional assumptions with S-matrix principles and with phenomenology. Naturally it would be even more valuable to prove in S-matrix theory some property of the amplitude, which allows to estimate the remaining error.
Up to this point, the much emphasized contact to phenomenology of the above mentioned results is of a principle nature. The numerical comparison with experimental data of crossing sumrule and dispersion relation needs, due to the approximation with Legendre functions, which is involved, very tedious computations. As a first attempt a numerical bound for high energy $\pi\pi$-scattering will be calculated in collaboration with A. DE LA TORRE, using the dispersion relation and elstic t-channel unitarity to bound s-channel partial waves with large angular momenta.
Many other applications can be thought of, e.g. the determination of the profile function for high values of the impact parameter b in p—π and p—p scattering. This would involve the use of coupled t-channel unitarity equation. We should be glad if the ideas and methods, described in this work, will be of use for such problems in the future.

II. Basic Assumptions

In this section we state the basic properties of the amplitude, which are needed for the later considerations. They follow from analyticity, crossing and unitarity, as well as from additional assumptions. For the sake of simplicity, the scattering of neutral pseudoscalar mesons will be treated. The extension to real π-meson with isospin one will be given in ref. [46], for an application of the dispersion relation. The mass unit is chosen as $m_\pi = 1$. The usual Mandelstam variables are, expressed in terms of center of mass energy E_{cm}, momentum q_{cm} and scattering angel Θ:

$$s = E_{\text{cm}}^2 = 4(1 + q_{\text{cm}}^2)$$
$$t = -2q_{\text{cm}}^2(1 - \cos\Theta)$$
$$u = -2q_{\text{cm}}^2(1 + \cos\Theta)$$
$$s + t + u = 4.$$

i) Analycity properties

The elastic scattering amplitude is the boundary value of an analytic function $F(s, t)$. The full domain of analyticity of $F(s, t)$ as function of two complex variables s and t is not yet known. In the framework of axiomatic field theory it has been possible to show, that the following regions belong to the analyticity domain of $F(s, t)$: For fixed $s > 0$ the Lehmann-Martin ellipse [12] E_4 in the complex t-plane

$$E_4 = t \in C: \left|Z + \sqrt{Z^2 - 1}\right| \leq Z_4 + \sqrt{Z_4{}^2 - 1}, \qquad Z = 1 + \frac{2t}{2 - 4}, \; Z_4 = 1 + \frac{8}{s - 4}; \tag{II.1}$$

for fixed t, $|t| < 4$, the complex s-plane except for the cuts $s \in [4, \infty)$; $s \in (-\infty, -t]$. The physical domain of the s-channel is reached for values of $s \geqq 4$ above the cut and t between $4 - s$ and 0. The analyticity domain of $F(s, t)$ as function of two complex variables contains a path which links this s-channel region with values $t \geqq 4$, $s \leqq 0$. This is important for the crossing property, discussed below. The derivation of these results and references to the original papers, as well as a more complete discussion of the axiomatic analyticity domain can be found in [5] and in the review of SOMMER [18].
The boundary of the analyticity domain as given by (II.1) is not dictated by physical considerations but reflects present time's possibilities to analyse the causality structure of field theory. It is very well possible, that the full analyticity domain of $F(s, t)$ is much larger, bounded only by the positions of singularities, which unitarity and crossing produce. Therefore it makes sense to look for consequences of this larger analyticity domain. This program was initiated by MANDELSTAM [13], who determined the analycity domain from the Feynman graphs of perturbation theory. Later on this idea of maximal analyticity became an axiom in the analytic S-matrix theory [19], which avoids the notion of fields.
Mandelstam analycity for the elastic amplitude means: $F(s, t)$ is analytic in the product of the s-plane, cut for $s \geqq 4$, $s \leqq -t$ and the t-plane cut for $t \geqq 4$, $t \leqq -s$.
The absorptive part in the s-channel:

$$A(s, t) = \frac{1}{2i}\,[F(s + i0, t) - F(s - i0, t)], \tag{II.2}$$

is analytic for fixed $s \geqq 4$ in the complex t-plane, cut for $t \geqq \sigma(s)$, $4 - t - s \geqq \sigma(s)$. The line $t = \sigma(s)$ in the s, t plane denotes the boundary of the domain, in which $A(s, t)$ develops an imaginary part $\varrho(s, t)$, which again can be expressed as a discontinuity:

$$\varrho(s, t) = \frac{1}{2i}\,[A(s, t + i0) - A(s, t - i0)], \qquad t \geqq \sigma(s). \tag{II.3}$$

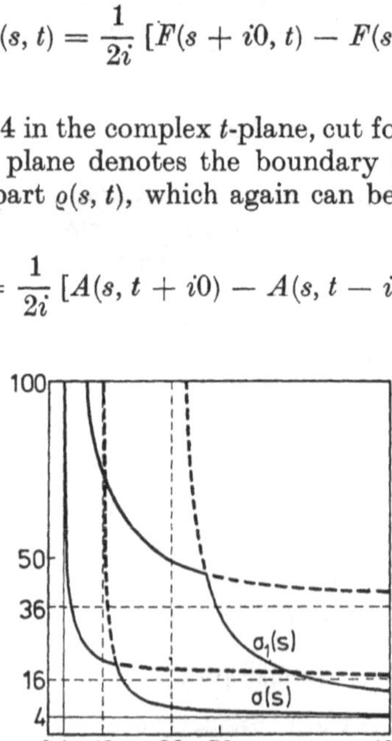

Fig. 1. The boundary of support of the Mandelstam double spectral function $\varrho(s, t)$. For $t \leqq \sigma_1(s)$, $\varrho(s, t)$ is positive (cf. VI.20)

This boundary is given by

$$\sigma(s) = \begin{cases} \dfrac{16s}{s-4} & s \leqq 20 \\[2ex] \dfrac{4s}{s-16} & s > 20\,. \end{cases} \tag{II.4}$$

The line $t = \sigma(s)$ in the s, t plane is symmetric with respect of permutation of s and t, $t = \sigma(s)$ is equivalent to $s = \sigma(t)$ (cf. fig. 1).
Sometimes in the following not the full Mandelstam analyticity will be used, but for fixed $s \geqq 4$ analyticity of $A(s, t)$ in the ellipse $E_{\bar t}$ with right extremity $\bar t$, which has t- and u-channel cuts, for $s \geqq \sigma(t)$ (cf. fig. 2):

$$E_{\bar t} = \Big\{t \in : \left|z = \sqrt{z^2-1}\right| \leqq \bar z + \sqrt{\bar z^2 - 1}\,; \qquad z = 1 + \frac{2t}{s-4}, \quad \bar z = 1 + \frac{2\bar t}{s-4},$$
$$t \neq \sigma(s) + \lambda, \qquad 4 - s - t \neq \sigma(s) + \lambda, \qquad \lambda \geqq 0\Big\}, \tag{II.5}$$

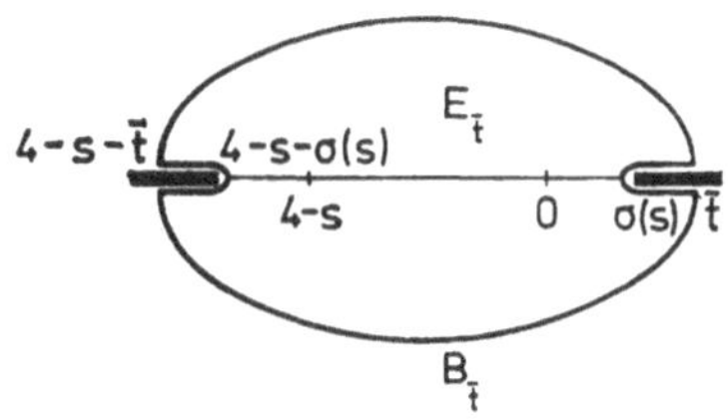

Fig. 2. The analyticity domain $E_{\bar t}$ (II.5) in the complex t-plane

ii) Crossing

As stated above, the axiomatic analyticity domain of $F(s, t)$ contains paths, which connect every point of the physical s-channel region with every point of the t-channel, i.e. $t \geqq 4$ avove the cut, $4 - t \leqq s \leqq 0$, and to every point of the u-channel, i.e. $u \geqq 4$ above the cut, $4 - u \leqq s \leqq 0$. This is the precise statement of the crossing property [20]. Since $F(s, t)$ is the amplitude for the scattering of identical mesons, this implies the symmetry under permutations of the variables:

$$F(s, t) = F(t, s) = F(s, u)\,. \tag{II.6}$$

If Mandelstam analyticity is assumed, the double spectral function $\varrho(s, t)$ is symmetric due to crossing:

$$\varrho(s, t) = \varrho(t, s) \tag{II.7}$$

iii) Unitarity

The implications of unitarity are most easily formulated for partial waves. The partial wave expansion for $F(s, t)$ is defined by

$$F(s, t) = \sum_{\text{even}\, l} (2l + 1)\, f_l(s)\, P_l\left(1 + \frac{2t}{s-4}\right), \qquad t \in E_4\,. \tag{II.8}$$

It converges for $t \in E_z$ due to the analyticity of $F(s, t)$ in this domain. $t - u$ symmetry implies, that only even values of l appear in the sum (II.8). For these partial waves $f_l(s)$

the following unitarity properties hold

$$\operatorname{Im} f_l(s) = \varrho(s)\, |f_l\,(s)|^2 + \frac{\sqrt{s}}{s-4}\, \eta_l(s), \qquad l = 0, 2, 4, \ldots, \tag{II.9}$$

$$\varrho(s) = \frac{\sqrt{s-4}}{s}. \tag{II.10}$$

$\eta_l(s)$ are the contributions of production amplitudes in the unitarity relation for the elastic amplitude. At this point part of the information implied by S-matrix principles is lost, since no use will be made of the complicated set of unitarity relations involving $\eta_l(s)$. Fortunately eq. (II.9) implies the bound

$$\eta_l(s) \leqq \frac{1}{4} - \frac{s-4}{s} \left(\operatorname{Re} f_l(s)\right)^2 \leqq \frac{1}{4}. \tag{II.11}$$

Furthermore the inelasticies $\eta_l(s)$ only contribute for $s \geqq 16$ and are positive:

$$\eta_l(s) \geqq 0, \qquad \eta_l(s) = 0 \quad \text{for} \quad s \leqq 16. \tag{II.12}$$

Without further need of inelastic amplitudes, unitarity therefore leads to the properties for the elastic amplitude alone:

$$\begin{aligned} &\operatorname{Im} f_l(s) = \varrho(s)\, |f_l(s)|^2 && 4 \leqq s \leqq 16 \\ &0 \leqq \operatorname{Im} f_l(s) \leqq 1/\varrho(s) && 4 \leqq s. \end{aligned} \tag{II.13}$$

This information from direct channel unitarity is one essential ingredient to the derivation of all bounds and relations for the elastic amplitude [*6*, *7*].

If Mandelstam analyticity is assumed, elastic unitarity in the crossed t-channel leads to an additional restriction. The double spectral function satisfies the Mandelstam equation [*13*]

$$\varrho(s, t) = \frac{4}{\pi \sqrt{t-4}} \int_D ds_1\, ds_2 K^{-1/2} A(s_1, t)\, A^*(s_2, t) \tag{II.14}$$

where

$$K = K(s, s_1, s_2, t) = s^2 + s_1{}^2 + s_2{}^2 - 2(ss_1 + ss_2 + s_1 s_2) - \frac{s_1 s_2 s}{t/4 - 1} \tag{II.15}$$

and the integration domain is given by

$$D = \{s_1, s_1 : s_1 \geqq 4, \quad s_2 \geqq 4, \quad K(s, s_1, s_2, t) \geqq 0\}. \tag{II.16}$$

iv) Polynomial boundedness

It follows from axiomatic field theory, that $F(s, t)$ is polynomial bounded in s, for $t \in E_4$ [*21*]:

$$|F(s, t)| \leqq F s^{N\,1)}, \qquad t \in E_4 \tag{II.17}$$

By use of s-channel unitarity and analyticity, JIN and MARTIN were able [*22*] to show that $N \leqq 2$ for $|t| < 4$. In fact the stronger bound holds [*23*] for $0 \leqq t \leqq 4$

$$N < 1 + \frac{\sqrt{t/4}}{4} + \varepsilon, \varepsilon > 0, \qquad 0 \leqq t \leqq 4. \tag{II.18}$$

[1]) Since $F(s, t)$ is a distribution in field theory, the bound (II.17) must be understood in the distribution theoretical sense: $\int_4^\infty ds\, F(s, t)\, h(s)$ exists for $|h(s)| \leqq s^{-N-1-\varepsilon}$, $\varepsilon > 0$.

From (II.17) and (II.18) follows the validity of the Froissart-Gribov representation for the t-channel partial waves with angular momenta $J \geqq 1 + \sqrt{t/4} + \varepsilon$ for $0 \leqq t < 4$.

$$f_J(t) = \frac{4}{\pi(4-t)} \int_4^\infty ds Q_J\left(1 + \frac{2s}{4-t}\right) A(s, t), \qquad 0 \leqq t \leqq 4. \tag{II.19}$$

The bound on the amplitude (II.17) and its integrated form (II.19) is the second essential ingredient for the proof of high energy theorems [6], since with its help the contribution of s-channel partial waves with high angular momenta can be estimated.
In order to exploit the additional assumption of Mandelstam analyticity, a similar boundedness property of $A(s, t)$ for t outside the Lehmann-Martin ellipse must hold.
A proof for polynomial boundedness in this domain starting from S-matrix principles is not yet available. However, all models, which are used for a description of strong interactions as e.g. potential scattering [24], multipheripheral model [25], dual model [26], lead to amplitudes which are polynomial bounded. Therefore the following assumption is introduced:

$$|A(s, t)| \leqq A \cdot s^{N(t)}, \qquad t \in E_i, \tag{II.20}$$

where the power $N(t)$ may change with t, such that rising Regge trajectories are covered by (II.20), too.
Now the Froissart-Gribow representation for the reduced partial waves $\tilde{f}_J(t)$ can be continued in the whole domain E_i, for $J > N(t)$, due to the analyticity of $A(s, t)$ in this domain and the bound (II.20):

$$\tilde{f}_J(t) = \left(\frac{4}{t-4}\right)^J f_J(t) = \frac{1}{\pi}\left(\frac{4}{t-4}\right)^{J+1} \int_4^\infty ds Q_J\left(1 + \frac{2t}{s-4}\right) A(s, t), \qquad J > N(t), \qquad t \in E_i. \tag{II.21}$$

Let me summarize the enumeration of basic properties, which will allow to exploit t-channel structures for s-channel partial waves with high angular momenta. Essentially in two direction assumptions have to be added to the axiomatic properties of the amplitude. In order to reach the t-channel directly for large s, the larger analyticity domain E_i (II.5) is needed, on the other hand the magnitude of $A(s, t)$ has to be bounded in the sense of (II.20) in E_i, if consequences of unitarity in the crossed channel are to be drawn, this will become clear in the following sections. It is important to notice, that both additional assumed properties hold in all current realistic models for the description of strong interactions. Therefore it is not an unrealistic hope, that the true amplitude shares these properties.

III. A Scaling Law for s-Channel Partial Waves

This scaling law describes the consequence of t-channel unitarity at threshold for s-channel partial waves. Originally it had been suggested as a sufficient condition on s-channel partial waves, which guaranties via the Froissart-Gribov representation (II.19) the correct threshold behaviour at $t = 4$ for the t-channel partial waves $f_J(t)$, $J = 2, 4, \ldots$ [27]. Due to unitarity (II.9), this behaviour near $t = 4$ for the reduced partial waves

$\tilde{f}_J(t) = \left(\frac{t-4}{4}\right)^{-J} f_J(t)$ is the following:

$$\tilde{f}_J(t) = a_J + a_J^{(1)}(t-4) + \cdots \left(\frac{4-t}{4}\right)^{J+1/2} \frac{2}{\sqrt{t}} \left(a_J^2 + 2a_J a_J^{(1)}(t-4) + \cdots\right) \quad \text{(III.1)}$$

Therefore, the $(J+1)$ derivative of $\tilde{f}_J(t)$ diverges for $t \to 4$:

$$\frac{\partial^{J+1}}{\partial t^{J+1}} \tilde{f}_J(t) = \frac{1}{2} \frac{\Gamma(J+3/2)}{4^J \sqrt{\pi}} a_J^2 (4-t)^{-1/2}. \quad \text{(III.2)}$$

The Froissart-Gribov representation (II.19), valid for $t \leqq 4$, $J \geqq 2$, relates this t-channel property (III.2) to the s-channel absorptive part $A(s, t)$ and, due to its partial wave expansion (II.8) to the s-channel partial waves:

$$\sqrt{\pi} \frac{\Gamma(J+1)}{\Gamma(J+3/2)} \frac{\partial^{J+1}}{\partial t^{J+1}} \int_4^\infty ds s^{-J-1} \sum_l (2l+1) \operatorname{Im} f_l(s) P_l\left(1 + \frac{2t}{s-4}\right)$$

$$\xrightarrow[t \to 4]{} \frac{\sqrt{\pi}}{2} \frac{\Gamma(J+3/2)}{4^J} a_J^2 (4-t)^{-1/2}. \quad \text{(III.3)}$$

It has been shown in ref. [27], that (III.3) is fullfilled, if the s-channel partial waves $\operatorname{Im} f_l(s)$ obey the following scaling law[2]) in the limit
$s \to \infty, l \to \infty, x = l \cdot s^{-3/2}$:

$$\operatorname{Im} f_l(s) = \sqrt{2\pi} \frac{1}{s^3} \frac{\beta(x)}{\sqrt{x}} e^{-16/3x} e^{-4sx}. \quad \text{(III.4)}$$

The moments of the structure function $\beta(x)$ are given by integrals over $A(s, 4)$, as can be seen using (III.3) and the Froissart-Gribov representation for the scattering lengths a_J:

$$\int_0^\infty dx x^{J+1/2} \beta(x) = \frac{1}{\sqrt{2\pi}} \frac{\Gamma(J+3/2)^2}{\Gamma(J+1)} \frac{1}{2^J} a_J^2$$

$$= \frac{1}{\sqrt{2\pi^2}} 2^{-J} \Gamma(J+1) \int_4^\infty ds s^{-J-1} A(s, 4)^2, \qquad J = 2, 4, \ldots \quad \text{(III.5)}$$

Later on COMMON [28] showed, that the moment problem (III.5) can be uniquely solved, leading to the following representation:

$$\beta(x) = \frac{4}{\sqrt{2\pi^3} \sqrt{x}} \int_4^\infty ds_1 \int_4^\infty ds_2 e^{-2xs_1s_2} A(s_1, 4) A(s_2, 4). \quad \text{(III.6)}$$

In this derivation of the scaling law (III.4), (III.6) use is made of axiomatic properties only, in particular of analyticity in the Lehmann-Martin ellipse E_4. The scaling law for s-channel partial waves is no necessary consequence of these properties alone, it only gives a sufficient condition to produce the correct t-channel singularity (III.2) at threshold. Furthermore the considerations inside the Lehmann-Martin ellipse do not allow

---

[2]) The factor $e^{-16/3x}$ is missing in the scaling law of ref. [27] due to the use of an incorrect asymptotic form for $P_l(1 + 2t/(s-4))$. This has been observed by COMMON [31], see below. The correct behaviour of P_l will be discussed in appendix A, cf. eq. (A. 4).

statements on the domain of validity of the scaling law. The formulae (III.4), (III.6) are consistent with the s-channel unitarity bound (II.13) for

$$l \geqq \frac{\sqrt{s}}{2}\left(\alpha(4) - 1\right) \ln s \tag{III.7}$$

where $\alpha(4)$ is the position at $t = 4$ of the leading Regge singularity [27]. A rigorous proof for the scaling law and statements on the domain of its validity need analyticity and polynomial boundedness of $A(s, t)$ outside the axiomatic domain, as will be shown below. If $A(s, t)$ is assumed to be analytic in the domain $E_{\bar{t}}$ (II.5), Froissart-Gribov like representations for the imaginary parts of s-channel partial waves hold. This can be seen by using the Neumann formula

$$P_l(z) = -\frac{1}{\pi i}\left(Q_l(z + i0) - Q_l(z - i0)\right), \; -1 \leqq z \leqq 1 \tag{III.8}$$

in the definition for $\operatorname{Im} f_l(s)$, viz.

$$\operatorname{Im} f_l(s) = \frac{1}{2(s-4)} \int\limits_{4-s}^{0} dt P_l\left(1 + \frac{2t}{s-4}\right) A(s, t) \tag{III.9}$$

and by distorting the integration contour to the boundary $B_{\bar{t}}$ of the ellipse $E_{\bar{t}}$ in the complex t-plane (cf. fig. 2):

$$\begin{aligned} \operatorname{Im} f_l(s) &= \Theta\left(s - \sigma(\bar{t})\right) \frac{4}{\pi(s-4)} \int\limits_{\sigma(s)}^{\bar{t}} dt Q_l\left(1 + \frac{2t}{s-4}\right) \varrho(s, t) \\ &\quad + \frac{2}{i\pi(s-4)} \int\limits_{B_{\bar{t}}^+} dt Q_l\left(1 + \frac{2t}{s-4}\right) A(s, t) \\ &= \alpha(s, l) + \varepsilon(s, l). \end{aligned} \tag{III.10}$$

It is due to $t - u$ symmetry, that in (III.10) only the contribution of the right cut and the integral on the right half $B_{\bar{t}}^+$ (cf. fig. 2) of the boundary of $E_{\bar{t}}$ appears.
For values $\bar{t} \leqq 16\, \varrho(s, t)$ can be expressed in terms of $A(s, t)$ by means of the Mandelstam equation (II.14), yielding

$$\begin{aligned} \alpha(s, l) &= \Theta\left(s - \sigma(\bar{t})\right) \frac{4}{\pi(s-4)} \int\limits_{\sigma(s)}^{\bar{t}} dt Q_l\left(1 + \frac{2t}{s-4}\right) \frac{4}{\pi\sqrt{t-4}} \\ &\quad \times \int\limits_D ds_1\, ds_2 K(s, s_1, s_2, t)^{-1/2} A(s_1, t) A^*(s_2, t). \end{aligned} \tag{III.11}$$

It now depends on the assumed magnitude of the analyticity domain $E_{\bar{t}}$ and on the assumed degree of polynomial boundedness, for which values of l (III.10) and (III.11) rigorously imply the scaling law.
Yndurain [29], as well as Grassberger and Martin [30], choose assumptions as weak as possible. They show, that it is sufficient to have analyticity in $E_{\bar{t}}$, where

$$\bar{t}(s) = \frac{4\sqrt{s} + 8}{\sqrt{s} - 6} \to 4 + \frac{32}{\sqrt{s}} \quad \text{for} \quad s \to \infty \tag{III.12}$$

and to assume in $E_{\bar{t}}$ the bound

$$|A(s, t)| \leqq A \cdot S^N \tag{III.13}$$

for any finite N, in order to prove the scaling law for l-values

$$l \sim s^{1+\delta}, \qquad 1/2 \geqq \delta > 0, \qquad s \to \infty. \tag{III.14}$$

We will reproduce the proof of Ynduráin [29], taking into account some corrections concerning the asymptotic properties of the Legendre functions, which are due to Common [31]. The contribution $\varepsilon(s, l)$ of (III.10), coming from the integration along the boundary $B_{\bar{t}}^+$ of $E_{\bar{t}}$ is for $s \to \infty$ bounded by

$$|\varepsilon(s, l)| \leqq \frac{A \cdot r}{\sqrt{2} \cdot \pi} s^{(N-1)/2} \frac{1}{\sqrt{x}} \mathrm{e}^{-4sx-16\sqrt{s}\cdot x+80/3x} \tag{III.15}$$

where $x = l \cdot s^{-3/2}$, $l \leqq s^{1+\delta}$, $r \cdot s$ denotes the lengths of $B_{\bar{t}}^+$. Here use has been made of the property that $|Q_l|$ reaches its maximum on $B_{\bar{t}}$ at the right extremity:

$$\left|Q_l\left(1 + \frac{2t}{s-4}\right)\right| \leqq Q_l\left(1 + \frac{2\bar{t}}{s-4}\right) \quad \text{for} \quad t \in B_{\bar{t}'}, \tag{III.16}$$

as well as of the asymptotic behaviour (A.6) of $Q_l\big(1 + 2\bar{t}(s)/(s-4)\big)$, and of the bound (III.13) on $A(s, t)$. Notice the factor $s^{(N-1)/2}\, e^{-16\sqrt{s}\cdot x}$ in the bound (III.15), which implies that $\varepsilon(s, l)$ vanishes for $x > s^{-1/2+\delta}$, $\delta > 0$, faster than any power of $1/s$ in comparison to the scaling law (III.4.)

It remains to show that the first part of (III.10), $\alpha(s, l)$, reproduces the scaling law. In the asymptotic region $s \to \infty$ the Kibble function in (III.11) reduces to

$$K(s, s_2, s_2, t) \to s\left(s - \frac{s_1 s_2}{t/4 - 1}\right), \tag{III.17}$$

the integration domain D therefore is given by

$$D = \left\{s_1, s_2 : s_1 \geqq 4,\ s_2 \geqq 4,\ s_1 \cdot s_2 \leqq \big((t/4) - 1\big) \cdot s\right\}. \tag{III.18}$$

In this integration domain s_1 and s_2 stay below $2\sqrt{s}$ for $t < \bar{t}(s)$ (cf. III.12). Therefore in D

$$t \leqq 4 + \frac{32}{\sqrt{s}} \leqq 4 + \frac{64}{s_1 - 16} \tag{III.19}$$

which lies outside the support of $\varrho(s_1, t)$ (cf. (II.4) and fig. 1). Consequently $A(s_1, t)$, and similarly $A(s_2, t)$, are positive in the integration region of (III.11) and the order of integration may be interchanged. This leads to

$$\alpha(s, l) = \frac{4Q_l\left(1 + \frac{2\sigma(s)}{s-4}\right)}{\pi^2 \sqrt{s}\,(s-4)} \int\limits_{s_1, s_2 \geqq 4}^{s_1, s_2 \leqq (\bar{t}(s)/4-1)\cdot s} ds_1 ds_2 A(s_1, t')\, A(s_2, t')\, 4 \int\limits_{4+4s_1s_2/s}^{\bar{t}(s)} dt \frac{Q_l\left(1 + \frac{2t}{s-4}\right)}{Q_l\left(1 + \frac{2\sigma(s)}{s-4}\right)}$$

$$\times \frac{1}{\sqrt{s(t-4) - 4s_1, s_2}} \tag{III.20}$$

where $t' \in [\sigma(s), \bar{t}(s)]$. With the help of the asymptotic formula for the Legendre function (A.9) and (A.10) the t-integration can be performed in the limit $s \to \infty$:

$$\alpha(s, l) = \frac{4}{\pi} \frac{1}{s^3} \frac{1}{x} e^{-4sx-16/3x} \int_D ds_1 \int ds_2 e^{-2xs_1s_2} A(s_1, t') A(s_2, t'). \tag{III.21}$$

As a last step the limit $A(s_1, t') \to A(s_1, 4)$ for $s \to \infty$ has to be performed under the integral in eq. (III.21), which is allowed due to the absolute convergence of the integral. Thus the validity of the scaling law under assumptions (III.12), (III.13) is established.
The values of l (III.14) for which the assumptions (III.12), (III.13) imply the scaling law are too high to be of phenomenological interest. Stronger assumptions lead to a proof of the scaling law for smaller values of l. A generalization of Yndurains proof is valid for

$$l \sim \sqrt{s}\,(\ln s)^{1+\delta}, \qquad \delta > 0 \qquad s \to \infty, \tag{III.22}$$

if the following assumptions are made: $A(s, t)$ is analytic in the domain $E_{\bar{t}}$ (II.5), where

$$\bar{t} = 4 + \varepsilon, \qquad \varepsilon > 0, \tag{III.23}$$

$A(s, t)$ is bounded by

$$|A(s, t)| \leqq s^N \quad \text{for} \quad t \in E_{\bar{t}}, \tag{III.24}$$

$$|A(s, t)| \leqq A(s, 4)\, s^{\gamma(t-4)}, \qquad \gamma > 0, \qquad 4 \leqq t \leqq \bar{t}. \tag{III.25}$$

Assumptions (III.23)—(III.25) lead to the estimate

$$|\varepsilon(s, l)| \leqq \text{const}\; e^{-4l/\sqrt{s}} s^{N-\varepsilon/2(\ln s)^\delta}, \tag{III.26}$$

which for $s \to \infty$ vanishes faster than the scaling law. The integral (III.11) for $\alpha(s, l)$ gets its contributions effectively from the interval $t \leqq 4 + \text{const} \cdot (\ln s)^{-1-\delta/2}$ only, due to the bound (III.25). For these values of t, $A(s, t)$ may be replaced by $A(s_1, 4)$ in the integral (III.21) and the scaling law follows.
If even stronger assumptions are imposed, especially a smoothness property

$$|A(s, t) - s^{\alpha'(t-4)} A(s, 4)| \leqq s^{\alpha'(t-4)} A(s, 4)\, (s^{\gamma(t-4)^{1+\varepsilon}} - 1), \tag{III.27}$$

a modified scaling law is valid for values

$$l = \lambda \sqrt{s}\, \ln s. \tag{III.28}$$

The modification consists in replacing the argument x of the structure function $\beta(x)$ by $x - 2\alpha' \ln s/s$. For the details of the derivation see the original paper [*32*].
Condition (III.27) is fullfilled e.g. in Regge models with differentiable trajectories. It is therefore meaningfull to compare the scaling law for values (III.28) of l with experimental data. Unfortunately, the absorptive part $A(s, 4)$, which enters $\beta(x)$ (cf. eq. III.6) is not directly measurable. Therefore the simple Ansatz for the πN scattering amplitude at $t = 4$,

$$A(s, 4) = A \cdot s^{1+\varepsilon}, \tag{III.29}$$

with small ε in view of the small slope of the pomeron trajectory will be used to calculate the dependence of the partial waves of NN scattering on $b = 2l/\sqrt{s}$ via the scaling law.

This is compared with the b-behaviour at ISR energies of the measured NN profile function [33] in fig. 3. With one adjusted parameter, namely the magnitude A of the absorptive part of the πN amplitude at t-channel threshold (eq. (III.29), the b dependence for $b > 1{,}4$ fm is correctly reproduced. This shows, that for these values of b and s, t-channel structure at threshold determines the s-channel partial waves.

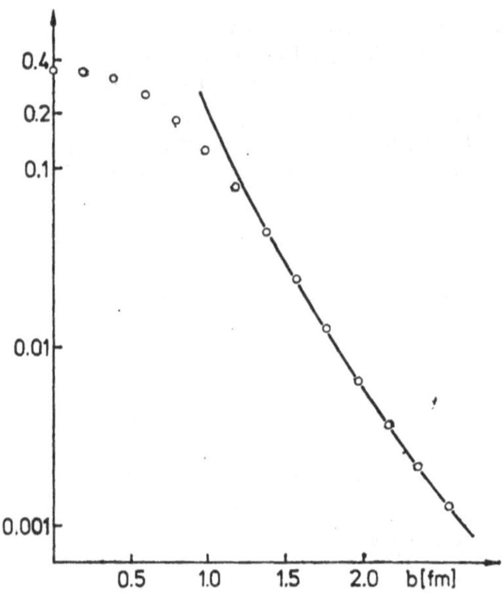

Fig. 3. Comparison of the scaling law (III.4) (solid line) with experimentally measured partial waves [33] (circles) at $p_{lab} = 1480$ GeV

IV. Approximate Crossing Sumrules for Partial Waves

The scaling law of the last section is an asymptotic property of partial waves. Comparison with experiment always involves finite values of s and l, and it is not known a priori, where the asymptotic region starts. In this section we will describe the derivation of sumrules [34] for weighted integrals over s-channel partial waves, which avoid this difficulty. Moreover, partial waves with phenomenologically interesting values of angular momentum $l = \sqrt{s}/2(b_0 + \lambda \ln s)$ will be involved without specific additional assumptions on $A(s, t)$ near $t = 4$ like (III.25), (III.27).
However, the basic properties analyticity and polynomial boundedness of $A(s, t)$ in an larger region than the axiomatic domain will be used:

i) Analyticity in the region $E_{\bar{t}}$ (of II.5 and fig. 2), with $\bar{t} > 4$
ii) Polynomial boundedness in s for $t \in E_{\bar{t}}$ (cf. II.20)

is assumed in the form

$$|A(s, t)| \leqq A s^{N(t)}, \qquad 4 \leqq t \leqq \bar{t} \tag{IV.1}$$

$$A(s, t) \leqq A s^{N(\bar{t})} \left(\frac{\bar{t}}{t}\right)^2 \quad \text{for} \quad t \in B_{\bar{t}}. \tag{IV.2}$$

Here again $B_{\bar t}$ is the boundary of the ellipse $E_{\bar t}$ (fig. 2):

$$B_{\bar t} = \left\{t \in C, \left|z + \sqrt{z^2 - 1}\right| = \bar z + \sqrt{\bar z^2 - 1},\ z = 1 + \frac{2t}{s-4},\ \bar z = 1 + \frac{2\bar t}{s-4}\right\}. \tag{IV.3}$$

From these assumptions, representation (III.10) for $\operatorname{Im} f_l(s)$ follows for all values of l. We consider the average of this relation:

$$\int\limits_{\sigma(\bar t)}^{\infty} ds\, c(s) \operatorname{Im} f_l(s) = \frac{4}{\pi} \int\limits_{\sigma(\bar t)}^{\infty} ds \frac{c(s)}{s-4} \int\limits_{\sigma(s)}^{\infty} dt\, Q_l\left(1 + \frac{2t}{s-4}\right) \varrho(s, t) + \int\limits_{\sigma(\bar t)}^{\infty} ds c(s)\, \varepsilon(s, l). \tag{IV.4}$$

In order to relate the integral over the doublespectral function to integrals over t-channel partial waves, the order of integration with respect to s and to t must be interchanged. This interchange will beallowed, if the s-integral converges absolutely, and this condition restricts the values of l, for which (IV.4) and the resulting sumrule can be used. Let α_0 and λ_0 be constants such that the power $N(t)$ in the polynomial bound (IV.1) obeys

$$N(t) \leqq \alpha_0 + \lambda_0 \sqrt{t}, \qquad 4 \leqq t \leqq \bar t \tag{IV.5}$$

and let furthermore the weight function $c(s)$ behave as

$$c(s) = c \cdot (\lambda \ln s)^{1/2}\, s^{-\alpha_0}\big(1 + O(1/s)\big), \quad \text{for } s \geqq \sigma(\bar t) \tag{IV.6}$$

then the integral over s in (IV.4) converges absolutely for $4 \leqq t \leqq \bar t$, if

$$l = l(s) = \frac{\sqrt{s}}{2}(b_0 + \lambda \ln s), \qquad b_0 \geqq 0, \qquad \lambda \geqq \lambda_0 \tag{IV.7}$$

where λ_0 in (IV.7) and α_0 in (IV.6) are restricted by (IV.5). This statement is a simple consequence of the asymptotic behaviour of $Q_l\big(1 + 2t/(s-4)\big)$ for these l-values (IV.7), (cf. A.11):

$$Q_l\left(1 + \frac{2t}{s-4}\right) \sim \text{const}\ s^{-\lambda\sqrt{t}}. \tag{IV.8}$$

We want to state another consequence of the restriction to these values (IV.7) of l: The boundedness by a constant of s-channel partial waves with these angular momenta follows immediately from the polynomial bound (IV.1) on the amplitude. For smaller angular momenta this boundedness of the partial waves can be proven only if more detailed information on the amplitude, e.g. its oscillations for $t \geqq 4$, is available. Relation (IV.4) can now be written as

$$\int\limits_{\sigma(\bar t)}^{\infty} ds\, c(s) \operatorname{Im} f_l(s) = \frac{1}{\pi} \int\limits_{4}^{\bar t} dt \int\limits_{\sigma(t)}^{\infty} ds\, \phi(s, t)\, \varrho(s, t) + \varepsilon \tag{IV.9}$$

$$\phi(s, t) = 4 \frac{c(s)}{s-4} Q_l\left(1 + \frac{2t}{s-4}\right). \tag{IV.10}$$

The reminder ε (which is related to contributions to $\operatorname{Im} f_l(s)$ from the boundary $B_{\bar t}$ of the ellipse $E_{\bar t}$ in the Froissart-Gribov integral (III.10)) can be estimated using (IV.6), (IV.7), assumption (IV.2) and the behaviour (A.10) of Legendre functions:

$$|\varepsilon| = \left| \int_{\sigma(\bar{t})}^{\infty} ds\, c(s)\, \varepsilon(s, l) \right| \leq \frac{2}{\pi} A \cdot c \cdot \Lambda \int_{\sigma(\bar{t})}^{\infty} ds\, Q_l \left(1 + \frac{2\bar{t}}{s-4}\right) \frac{s^{N(\bar{t})-\alpha_0}}{s-4} \left(1 + 0(1/s)\right)$$

$$\leq \text{const} \frac{\exp\left(-b_0 \sqrt{\bar{t}}\right)}{\alpha_0 + \lambda \sqrt{\bar{t}} - N(\bar{t})}, \qquad \text{(IV.11)}$$

where $\Lambda = \int_{B_{\bar{t}}^{+}} |dt|\, |\bar{t}/t|^2$.

From the inequality it is evident, that the contribution ε to the relation (IV.9) can be made arbitrarily small by choosing $b_0 \sqrt{\bar{t}}$ large enough.

The most important consequence of the choice (IV.7) for l and (IV.6) for $c(s)$ is the fact, that the weight $\phi(s, t)$ in (IV.9) is asymptotically related to the Legendre function $Q_J\left(1 + 2s/(t-4)\right)$, which projects from the amplitude the t-channel partial wave $f_J(t)$:

$$\frac{c(s)}{s-4} Q_{l(s)} \left(1 + \frac{2t}{s-4}\right) \to c \sqrt{\frac{\pi}{2\sqrt{t}}}\, e^{-b_0\sqrt{t}} s^{-1-\alpha_0-\lambda\sqrt{t}}$$

$$\to c \cdot \frac{1}{\sqrt{2}\sqrt{t}} \frac{\Gamma(J + 3/2)}{\Gamma(J+1)} \left(\frac{t-4}{4}\right)^{-J-1} e^{-b_0\sqrt{t}} Q_J \left(1 + \frac{2s}{t-4}\right)$$

for $s \to \infty$ (IV.12)

where

$$J = J(t) = \alpha_0 + \lambda \sqrt{t}. \qquad \text{(IV.13)}$$

This asymptotic crossing property of Legendre functions [35] suggests to replace the s-channel Legendre function $Q_l\left(1 + 2t/(s-4)\right)$ by the t-channel Legendre function $Q_J\left(1 + 2s/(t-4)\right)$ and to obtain in this way an approximate crossing relationship between the behaviour of the s-channel partial wave $\operatorname{Im} f_l(s)$ with angular momentum $l = \sqrt{s}/2$ $(b_0 + \lambda \ln s)$ and the t-channel partial wave $\operatorname{Im} f_J(t)$, $J = \alpha_0 + \lambda\sqrt{t}$. It can be argued that the approximation should be specially good for small values of α_0 and λ, which are still compatible with (IV.5), since then $J(t)$ is near the leading Regge singularity at $\alpha(t)$, $\operatorname{Re} \alpha(t) \leq N(t)$, which is produced by the large s behaviour of the amplitude, and for large s the connection (IV.12) between s-and t-channel Legendre function becomes exact. Actually, the error in the approximate relationship between s- and t-channel partial waves can be expressed explicitly in terms of t-channel partial waves with physical angular momenta. This is a consequence of a completeness property of the Legendre functions, which was derived in ref. [34]. Let us assume, for definitness, that $N(t)$ stays below 2 for $4 \leq t \leq \bar{t}$ (which, in phenomenological models, is true for $\bar{t} \leq m_f^2$), and α_0 and λ are chosen such, that

$$2 \geq J(t) = \alpha_0 + \lambda \sqrt{t} \geq N(t) \quad \text{for} \quad 4 \leq t \leq \bar{t}. \qquad \text{(IV.14)}$$

Then for every $t \in [4, \bar{t}]$ the set of reduced Legendre functions

$$\{\tilde{Q}_{J(t)}(s, t), \tilde{Q}_2(s, t), \tilde{Q}_4(s, t), \ldots\},$$

$$\tilde{Q}_J(s, t) = \left(\frac{t-4}{5}\right)^{-J-1} Q_J \left(1 + \frac{2s}{t-4}\right) \qquad \text{(IV.15)}$$

is complete in the space of functions $f(s)$, which decrease for $s \to \infty$ at least as $s^{-J(t)-1}$. The property of completeness implies in particular that every function of this class e.g. $\phi(s, t)$ of eq. (IV.10) can be arbitrarily good approximated by a linear combination of the set (IV.15):

$$\phi(s, t) = a_0^M(t)\, Q_{J(t)}(s, t) + \sum_{n=1}^{M} a_n^M(t)\, \tilde{Q}_{2n}(s, t) + \varepsilon^M(s, t) \tag{IV.16}$$

If the coefficients $a_n^M(t), n = 0, \ldots M$ are choosen appropriately the error $\varepsilon^M(s, t)$ of this approximation shrinks to 0, if the order M of approximation tends to infinity. The magnitude of the error can in particular be defined to be the norm of the function $\varepsilon^M(s, t)$ in a certain Hilbert space. Then the best coefficients $a_n^M(t)$ are uniquely determined by the requirement, that for a fixed order M this norm $\|\varepsilon^M(s, t)\|$ becomes minimal. The $a_n^M(t)$ are given by the solution of a set of $M + 1$ linear equations. The mathematical details of this approximation will be given in Appendix B, where also the proof for completeness of the set (IV.15) will be repeated.

Here it suffices that the approximation (IV.16) is valid with an error, which is as small as desired. Inserting (IV.16) into the relation (IV.9), the s-integration can be performed explicitely (cf. the Froissart-Gribov representation for the reduced t-channel partial waves $\tilde{f}_J(t)$, eq. (II.21)):

$$\int_{\sigma(\bar{t})}^{\infty} ds\, c(s)\, \mathrm{Im}\, f_l(s) = \int_4^{\bar{t}} dt \left\{ a_0^M(t)\, \mathrm{Im}\, \tilde{f}_{J(t)}(t) + \sum_{n=1}^{M} a_n^M(t)\, \mathrm{Im}\, \tilde{f}_{2n}(t) \right\} + \varepsilon^M + \varepsilon. \tag{IV.17}$$

In this crossing sumrule, the errors ε^M and ε can be made arbitrarily small, ε (cf. IV.11) by choosing $b_0\sqrt{\bar{t}}$ large enough, ε^M by taking a large enough number M of t-channel partial waves and thus making $\varepsilon^M(s, t)$ small:

$$|\varepsilon^M| \leqq A \cdot \frac{4}{\pi} \int_4^{\bar{t}} dt \int_{\sigma(t)}^{\infty} ds\, |\varepsilon^M(s, t)|\, s^{N(t)}, \tag{IV.18}$$

where the constant A comes from the bound (IV.1) on the amplitude.

As stated above, the high energy behaviour of the s-channel partial waves is restricted, if the values α_0 and λ, cf. eqs. (IV.6) and (IV.7), are choosen such that $J(t) = \alpha_0 + \lambda\sqrt{t}$ comes close to the leading Regge singularity $N(t)$, cf. (IV.1). The average over $\mathrm{Im}\, f_l(s)$ is then determined by the t-channel partial wave $\mathrm{Im}\, f_J(t)$ with the unphysical angular momentum $J = J(t)$, plus correction expressed in terms of physical t-channel partial waves. The advantage of having partial waves is, of course, that even for the unphysical angular momentum $J(t)$, the elastic unitarity restriction has the simple form

$$\mathrm{Im}\, \tilde{f}_J(t) = \left(\frac{t-4}{4}\right)^J \sqrt{\frac{t-4}{t}}\, |\tilde{f}_J(t)|^2, \tag{IV.19}$$

which can be used either to restrict the possible models for $\tilde{f}_J(t)$ or, in connection with the dispersion relation of the next section, to calculate $\tilde{f}_J(t)$ explicitely.

One drawback of the sumrule (IV.17) is that it is not possible to touch the highest Regge singularity in the t-channel, by choosing $J(t) = \alpha_0 + \lambda\sqrt{t} = N(t)$ for some value of t, since then the bound (IV.18) for ε^M diverges. Therefore, the consequences of a specific model for the Regge singularity cannot be exploited in this kind of approximation scheme.

A different approach of approximation in terms of Legendre functions [36] allows to avoid this difficulty. Now an integral rather than a discrete sum over values of $J \geqq J(t)$ is involved. As demonstrated in appendix B, the following relation between Legendre function holds, cf. (B.16)

$$\frac{c(s)}{s-4} Q_l \left(1 + \frac{2t}{s-4}\right) = \int_0^\infty dy a(t, y)\, Q_J \left(1 + \frac{2s}{t-4}\right) + \bar{\varepsilon}(s, t) \qquad \text{(IV.20)}$$

where

$$l + 1/2 = \sqrt{\frac{s}{2}}\, b, \qquad b = b_0 + \lambda \ln s, \qquad \text{(IV.21)}$$

$$J = \alpha_0 + \lambda \sqrt{t} \left(1 + \frac{(y-1)^2}{2y}\right), \qquad \text{(IV.22)}$$

$$c(s) = s^{-\alpha_0}\big(1 + O(1/s)\big) \qquad \text{(IV.23)}$$

$$a(y, t) = \frac{1}{2\sqrt{\pi}} \frac{1}{y} e^{-b_0\sqrt{t}(1+(y-1)/2y)^2} \frac{\Gamma(J+3/2)}{\Gamma(J+1)} \left(\frac{t-4}{4}\right)^{-J-1}. \qquad \text{(IV.24)}$$

The important property of the remainder $\bar{\varepsilon}(s, t)$ is its bound for large s, cf. (B.17):

$$\begin{aligned} \bar{\varepsilon}(s, t) &\leqq \left(c_1 \frac{b}{s} + c_2 \frac{t-4}{s} + c_3 \frac{b}{\sqrt{s}}\right) \cdot \frac{c(s)}{s-4} Q_l \left(1 + \frac{2t}{s-4}\right) \\ &\leqq \text{const} \frac{\log s}{\sqrt{s}} \cdot \frac{c(s)}{s-4} Q_l \left(1 + \frac{2t}{s-4}\right). \end{aligned} \qquad \text{(IV.25)}$$

If relation (IV.20), weighted with $\varrho(s, t)$, is integrated over the domain $4 \leqq t \leqq \bar{t}$, $\sigma(t) \leqq s \leqq \infty$, the following crossing sumrule for partical waves results:

$$\int_{\sigma(\bar{t})}^{\infty} ds c(s) \; \text{Im}\, f_l(s) = \int_4^{\bar{t}} dt \int_0^\infty d\xi a(\xi, t) \; \text{Im}\, f_J(t) + \bar{\varepsilon} + \varepsilon, \qquad \text{(IV.26)}$$

where l and J are defined in (IV.21) and (IV.22). The error ε is bounded like (IV.11), as in the sumrule (IV.17). The second remainder

$$\bar{\varepsilon} = \int_4^{\bar{t}} dt \int_{\sigma(t)}^{\infty} ds \int_0^\infty d\xi \frac{a(t, \xi)}{t-4} Q_J \left(1 + \frac{2s}{t-4}\right) \bar{\varepsilon}(s, t)\, \varrho(s, t) \qquad \text{(IV.27)}$$

can be continued to values of $\alpha_0 + \lambda\sqrt{t} \geqq N(t) - 1/2$, due to the bound (IV.1) on ϱ and the extra power of $s^{-1/2}$ in the bound (IV.25) on $\bar{\varepsilon}(s, t)$. Therefore, the contribution of the leading Regge singularity is fully contained in the t-channel partial waves, which are involved in (IV.26), and it is possible to study the consequences of a given singularity in $\text{Im}\, f_J(t)$ for s-channel partial waves, using the crossing sumrule (IV.26). The remainder $\bar{\varepsilon}$ can be treated analogous to the previous approximation scheme, and may be expressed in terms of physical t-channel partial waves with angular momenta $J = 2, 4, \ldots$, if $N(t) \leqq 3/2$.

As an application of the sumrule (IV.26), MÜTTER [37] chooses an explicit unitarity Ansatz for the Regge singularity of the t-channel partial waves near $t = 4$ and calculates via (IV.26) the behaviour at large s and large b_0 of s-channel partial waves $\text{Im}\, f_l(s)$, $= \sqrt{s}/2\, (b_0 + \lambda \ln s)$.

V. Dispersion Relation for Crossed Channel Partial Waves

As demonstrated in the last section, the t-channel partial waves with angular momenta $J \geqq \alpha_0 + \lambda\sqrt{t}$ close to the rightmost Regge singularity restrict via the sumrules (IV.17) and (IV.26) the high energy behaviour of s-channel partial waves with angular momenta $l = \sqrt{s}/2\,(b_0 + \lambda \ln s)$. It is therefore of great importance that a dispersion relation for these t-channel partial waves exists [41], in which substractions and contribution from the left hand cut are expressed in terms of measurable quantities.
In this respect the dispersion relation resembles the Roy equations [9] for partial wave amplitudes. However, the dispersion relation to be described in this section needs for its derivation the assumptions of analyticity of $A(s, t)$ in $E_{\bar{t}}$ (II.5) and boundedness of $A(s, t)$ in $E_{\bar{t}}$ (cf. II.20) according to (IV.1), (IV.2) whereas the Roy equations are valid on the basis of axiomatic analyticity (cf. section II). As a reward for the stronger assumptions, the new dispersion relation holds for unphysical partial waves with noninteger values of angular momentum $J = \alpha_0 + \lambda\sqrt{t}$ and is therefore relevant for the high energy properties in the crossed s-channel, in contrast to the Roy equations, which involve only physical partial waves with integer angular momenta, and which are used mainly to restrict the behaviour of the scattering amplitude at low and medium energies [10].
The derivation of the dispersion relation is based on a representation for $A(s, t)$, which has been applied since long, e.g. by Martin [42]. This representation follows from the use of the Froissart-Gribov formula (III.10) for partial waves $\operatorname{Im} f_l(s)$ with $l \geqq L(s)$,

$$L(s) = \left[\frac{\sqrt{s}}{2}\,(b_0 + \lambda \ln s)\right], \tag{V.1}$$

(where $[A]$ denotes the largest even integer smaller than A) in the partial wave expansion of $A(s, t)$ for values of $t < 4$:

$$\begin{aligned} A(s,t) &= \sum_{l=0}^{\infty} (2l+1) \operatorname{Im} f_l(s)\, P_l\left(1 + \frac{2t}{s-4}\right) \\ &= \Sigma_L(s,t) + \frac{2\cdot\Theta\big(s-\sigma(\bar{t})\big)}{\pi(s-4)} \sum_{l=L}^{\infty} (2l+1)\left(P_l\left(1+\frac{2t}{s-4}\right) + P_l\left(1+\frac{2u}{s-4}\right)\right) \\ &\quad\times \int_{\sigma(s)}^{\bar{t}} dt'\, Q_l\left(1+\frac{2t'}{s-4}\right) \varrho(s,t') + \varepsilon_L(s,t). \end{aligned} \tag{V.2}$$

Here Σ_L denotes the sum over the first $L-2$ waves

$$\Sigma_L(s,t) = \sum_{\substack{l=0 \\ l\,\text{even}}}^{L-2} (2l+1) \operatorname{Im} f_l(s)\; P_l\left(1 + \frac{2t}{2-4}\right), \tag{V.3}$$

$\Sigma_L(s, t)$ is a polynomial of degree $L-2$ in t.
The order of summation and integration in (IV.2) can be interchanged for $t < 4 < t'$, since for $t \leqq t' - \varepsilon$ the sum converges uniformly. An explicite espression for the sum of Legendre functions is [43]

$$\frac{4}{s-4} \sum_{l=L}^{\infty} (2l+1)\, P_l\left(1 + \frac{2t}{s-4}\right) Q_l\left(1 + \frac{2t'}{s-4}\right) = \frac{h_L(s,t,t')}{t'-t}, \tag{V.4}$$

$$h_L(s, t, t') = L \cdot \left\{ P_L\left(1 + \frac{2t}{s-4}\right) Q_{L-1}\left(1 + \frac{2t'}{s-4}\right) - P_{L-1}\left(1 + \frac{2t}{s-4}\right) Q_L\left(1 + \frac{2t'}{s-4}\right) \right\}. \tag{V.5}$$

The kernel $h_L(s, t, t')$ is an analytic function in t, and has the property (which will be derived in appendix C):

$$h_L(s, t, t') = 1. \tag{V.6}$$

With the help of (V.4) the representation for $A(s, t)$ is

$$A(s, t) = \sum_L(s, t) + \frac{\Theta(s - \sigma(\bar{t}))}{\pi} \int_{\sigma(t)}^{\bar{t}} dt' \left(\frac{h_L(s, t, t')}{t' - t} + \frac{h_L(s, u, t')}{t' - u} \right) \varrho(s, t') + \varepsilon_L(s, t) \tag{V.7}$$

$$\varepsilon_L(s, t) = \frac{1}{2\pi i} \int_{B_{\bar{t}}^+} \frac{dt'}{t' - t} h_L(s, t, t') \, A(s, t'). \tag{V.8}$$

Obviously the right hand side is analytic for $t \in E_{\bar{t}}$. Therefore, by analytic continuation, the representation (V.7) is valid in this domain. Due to (V.6) the discontinuity over the cut has the correct form, Im $A(s, t) = \varrho(s, t)$. Representation (V.7) can be viewed at as a L-times subtracted dispersion relation for $A(s, t)$, where the subtraction polynomial has been expressed in terms of physical s-channel partial waves with angular momenta $l \leqq L - 2$. The left hand cut is correctly incorporated in this relation, due to its $t - u$ symmetry. This representation, supplemented by the elastic unitarity equation (II.14) for $\varrho(s, t)$, could be used, according to Mandelstam's suggestion [*13*] as dynamical equations to determine $A(s, t)$. The difference here would be, that only s-channel partial waves with $l \geqq L(s)$ are determined by equations (V.7), (II.14) and those with $l \leqq L(s)$ are assumed to be known and are used as input, whereas Mandelstam and his many successors [*14*] tried to determine the full amplitude up to a substraction polynomial of small degree, corresponding to a Mandelstam representation with a small number of subtractions.

The reason for the restriction to higher values of l comes, as mentioned in the introduction, from the experience that the elastic part of the t-channel alone cannot determine the s-channel partial waves with low values of l, more information on inelastic t-channel contributions has to be used in order to calculate them. A second reason for the use fo the large $L(s)$ (cf. V.1) in the relation (V.7) is the following: The system of equations (V.7), (II.14) is extremely difficult to handle, only existence proofs are available for exact solutions [*15*, *16*], numerical treatment so far has to make drastic approximations [*14*], which are hard to control. It is just for values of $L(s)$ according to (V.1) that the approximation formalism of the last section is applicable. This leads to a relation among partial waves, for which elastic unitarity is expressed in the simple form (IV.19).

The reduced partial waves $\tilde{f}_J(t)$ are given according to (II.21) by the Froissart-Gribov projection of the representation (V.7):[3])

$$\tilde{f}_J(t) = g(t) + \frac{1}{\pi} \int_{\sigma(\bar{t})}^{\infty} ds \, \tilde{Q}_J(s, t) \, \frac{1}{\pi} \int_{\sigma(s)}^{\bar{t}} dt' \left(\frac{h_L(s, t, t')}{t' - t} + (t \to u) \right) \varrho(s, t') + \varepsilon(t), \tag{V.9}$$

---

[3]) We recall the definition of the reduced Legendre function $\tilde{Q}_J(s, t) = (t/4 - 1)^{-J-1} Q_J(1 + 2s/(t - 4))$.

$$g(t) = \frac{1}{\pi} \int_4^\infty ds\, \tilde{Q}_J(s,t) \qquad \sum_L(s,t), \tag{V.10}$$

where

$$J = J(t) = \alpha_0 + \lambda \sqrt{t} > N(t), \tag{V.11}$$

$$\varepsilon(t) = \frac{1}{\pi} \int_4^\infty ds\, \tilde{Q}_J(s,t)\, \varepsilon_L(s,t). \tag{V.12}$$

Just as in eq. (IV.4) of the last section it is necessary to interchange the order of s- and t-integration in (IV.9). The choice (V.1) for $L(s)$, together with (IV.11), makes the s-integration over $\varrho(s, t)$ absolutely convergent. This can be seen by inspecting the asymptotic behaviour of

$$\phi(s,t,t') = \tilde{Q}_J(s,t) \cdot \left(h_L(s,t,t') + \frac{t'-t}{t'-u} h_L(s,u,t')\right) \tag{V.13}$$

which is, according to appendix C,

$$\phi(s,t,t') = \frac{\Gamma(J+3/2)}{\Gamma(J+1)} \frac{\sqrt{t}+\sqrt{t'}}{2(tt')^{1/4}} e^{-b_0(\sqrt{t'}+\sqrt{t})} s^{-\alpha_0-\lambda\sqrt{t'}-1}\left(1+R(s,t,t')\right) \tag{V.14}$$

where $R(s, t, t') \to 0$ for $s \to \infty$. The assumed bound (IV.1) on $\varrho(s, t)$ guaranties with (V.14) the convergence of the s-integration. Since thus the change of order of integration is permitted, the approximation scheme of appendix B can be applied to express the weighted integral $\int ds\, \phi(s, t, t')\, \varrho(s, t')$ in terms of t-channel partial waves. We assume again that $N(t) \leqq 2$ for $4 \leqq t \leqq \bar{t}$, and choose λ, α_0 such that

$$2 \geqq J(t') = \alpha_0 + \lambda\sqrt{t'} \geqq N(t') \quad \text{for} \quad 4 \leqq t' \leqq \bar{t}. \tag{V.15}$$

To begin with, one calculates the coefficients $a_n{}^M(t, t')$, $n = 0, \ldots, M$ for the approximation

$$\phi(s,t,t') = a_0{}^M(t,t')\, \tilde{Q}_{J(t')}(s,t') + \sum_{n=1}^{M} a_n{}^M(t,t')\, \tilde{Q}_{2n}(s,t') + \varepsilon^M(s,t,t') \tag{V.16}$$

from the set of equations analogous to (B.9), in order to minimise the approximation error $\varepsilon^M(s, t, t')$.

Inserting this approximation (V.16) into the expression (V.9) for $\tilde{f}_J(t)$, the s-integration over the double spectral function can be explicitely performed (cf. the Froissart-Gribov representation (II.21)), resulting in the dispersion relation for $\tilde{f}_{J(t)}(t)$:

$$\begin{aligned}\tilde{f}_{J(t)}(t) &= \frac{1}{\pi} \int_4^{\bar{t}} \frac{dt'}{t'-t}\, a_0{}^M(t,t')\, \mathrm{Im}\, \tilde{f}_{J(t')}(t') \\ &+ \sum_{n=1}^{M} \frac{1}{\pi} \int_4^{\bar{t}} dt' \frac{a_n{}^M(t,t')}{t'-t}\, \mathrm{Im}\, \tilde{f}_{2n}(t') + g(t) + \varepsilon(t) + \varepsilon^M(t).\end{aligned} \tag{V.17}$$

There is no contribution to $\mathrm{Im}\, \tilde{f}_J(t)$ from the terms with $n \geqq 1$ in (V.17), since $a_n{}^M(t, t')$ vanishes for $t \to t'$, as $\phi(s, t, t) = \tilde{Q}_{J(t)}(s, t)$ due to the property (V.6) and definition

(V.13) of $\phi(s, t, t')$. The two remainders $\varepsilon(t)$ and $\varepsilon^M(t)$ can be estimated in the following way

$$|\varepsilon(t)| = \left| \frac{1}{\pi} \int_4^\infty ds\, \tilde{Q}_{J(t)}(s, t) \frac{1}{4\pi i} \int_{B_{\bar{t}}} \frac{dt'}{t' - t} h_L(s, t, t') A(s, t') \right|$$

$$\leqq \text{const.}\ A \cdot \Lambda(1 + R) \frac{1}{4\pi\Delta(t)} \frac{e^{-b_0(\sqrt{\bar{t}} - \sqrt{t})}}{\alpha_0 + \lambda\sqrt{\bar{t}} - N(\bar{t})} \frac{\Gamma(J + 3/2)}{\Gamma(J + 1)} \frac{\sqrt{t} + \sqrt{\bar{t}}}{2 \cdot (t \cdot \bar{t})^{1/4}}. \quad \text{(V.18)}$$

Here $\Delta(t) = \operatorname*{Min}_{t' \in B_{\bar{t}}} (t - t')$, Λ is defined as in (IV.11), and the bound (IV.2) on $A(s, t)$ has been used, as well as the asymptotic form (V.14) of $\phi(s, t, t')$.

$$|\varepsilon^M(t)| = \frac{1}{\pi^2} \left| \int_{\sigma(\bar{t})}^\infty ds \int_{\sigma(s)}^{\bar{t}} dt' \frac{\varepsilon^M(s, t, t')}{t' - t} \varrho(s, t') \right| \leqq A \cdot \frac{1}{\pi^2} \int_4^{\bar{t}} dt' \int_{\sigma(t)}^\infty ds\, s^{N(t')} \left| \frac{\varepsilon^M(s, t, t')}{t' - t} \right|. \quad \text{(V.19)}$$

The singularity at $t' = t$ in (V.19) is spurious, since $\varepsilon^M(s, t, t')$ vanishes for $t' = t$.
The bounds (V.18) and (V.19) show that for $t < \bar{t}$ both errors can be made arbitrarily small, for given bounds (IV.1), (IV.2) on $A(s, t)$: choosing $b_0\sqrt{\bar{t}}$ large makes $\varepsilon(t)$ small, a large order M of approximation in (V.16) deminishes $\varepsilon^M(s, t, t')$ and therefore $\varepsilon^M(t)$.
Equation (V.17) can be regarded as an approximate dispersion relation for $\bar{f}_J(t)$, in which substractions and contribution from the unphysical left hand cut are expressed in terms of physical partial waves: s-channel partial waves Im $f_l(s)$ with $l \leqq [\sqrt{s}/2\,(b_0 + \lambda \ln s)]$ contribute to $g(t)$ (cf. V.10), t-channel partial waves with even angular momenta smaller than 2 M appear as corrections to the dispersion integral. As has been seen in (V.18) and (V.19) the magnitude of the errors involved in the approximations depend on b_0, i.e. on how many s-channel partial waves are given as input, and on M, i.e. on how many t-channel partial waves are used.
Given these physical partial waves from experiment (or model), $\bar{f}_J(t)$, for $J(t) = \alpha_0 + \lambda\sqrt{t}$ near the highest Regge singularity, can be calculated from (V.17) and elastic unitarity (V.19), using well known N/D techniques [44]. Inserted into the crossing sumrule (IV.17) or (IV.26), $\bar{f}_J(t)$ helps to determine the s-channel partial waves with $l \geqq L(s)$.
Thus, dispersion relation (V.17) and crossing sumrule (IV.17), (IV.26) put strong constraints on the peripheral s-channel partial waves with $l \geqq L(s)$ in terms of the central partial waves with $l \leqq L(s)$.
In order to decide, how much of the s-channel partial wave at, say $b_0 = 1$ fm is determined by the near t-channel cut, e.g. $4 \leqq t \leqq \bar{t} = 1$ GeV, a numerical estimate of the errors ε, ε^M in (IV.17) and $\varepsilon(t)$, $\varepsilon^M(t)$ in (V.17) is indispensable. By inspection of the bounds (IV.11), (IV.18), (V.18), (V.19) on the errors, it is clear that it is the constant A and the power $N(t)$ in the bound (IV.1) and (IV.2) on $A(s, t)$, which have to be specified. Unfortunately, without further assumptions nothing can be said on the magnitude of A and $N(t)$. For the error discussion in practical applications some model for the magnitude of $A(s, t)$ has to be assumed, which allows numerical estimates of the errors.

VI. Bounds on the Approximation Errors

Let me recall the origin of the approximation errors. ε in the crossing sumrule (IV.17) and $\varepsilon(t)$ in the dispersion relation (V.17) can be expressed as integrals over the region $t \geqq \bar{t}$ of the double spectral function, if analyticity of $A(s, t)$ in the whole cut t-plane is assumed and the integration contour $B_{\bar{t}}^+$ in (III.10) is distorted to the cut $t \geqq \bar{t}$.

$$\varepsilon(s, l) = \frac{4}{\pi(s-4)} \int_{\bar{t}}^{\infty} dt\, Q\left(1 + \frac{2t}{s-4}\right) \varrho(s, t). \qquad \text{(VI.1)}$$

For high enough values of b_0 in the s-channel angular momentum $l = \sqrt{s/2}\,(b_0 + \lambda \ln s)$, ε and $\varepsilon(t)$ can be neglected, according to (IV.11) and (V.18), but the crucial question for applications is: are these contributions small for phenomenological interesting values of $b_0 \sim 1$ fm. The coincidence of the b-dependence of the scaling law which uses t-channel properties at $t = 4$ only, with the experimentally observed one for $b \geqq 1{,}4$ fm (cf. section III and fig. 3) could be an indication, that appreciably lower b values can be described by the contribution of an extended region $4 \leqq t \leqq \bar{t}$ in the t-plane.
The errors ε^M and $\varepsilon^M(t)$ on the other hand come from the approximation of these t-channel contributions by t-channel partial waves. Again these errors can be neglected, if a large enough number $M + 1$ of t-channel partial waves, which have to be known with a high enough precision, is considered in (IV.17) or (V.17). Here the typical approximation problem in encountered, that convergence of the approximation for $M \to \infty$ is achieved by strong cancellations of the individual terms in (IV.17) and (V.17), such that starting from some M, the uncertainties in the determination of Im $\tilde{f}_{2n}(t)$ will dominate. Extending the approximation in (IV.16) or (V.16) further to higher values of M will then become senseless. By choosing $J(t) = \alpha_0 + \lambda \sqrt{t}$ in (IV.18) and (V.17) near the leading Regge singularity, the contribution of the partial wave Im $\tilde{f}_{J(t)}$ is expected to dominate, and a low order M of approximation will give reasonable accuracy.
Of course, these qualitative discussions should be substanciated by numerical estimates of the errors in specific models. As an example we choose a Regge pole with trajectory $\alpha(t)$, which is bounded by $c_1\sqrt{t}$ for $t \to \infty$ and is very flat for $t \leqq \bar{t} = 50$ ($= 0{,}9$ GeV^2):

$$A(s, t) = A(t)\, s^{\alpha(t)} \qquad \text{(VI.2)}$$

$$\text{Re}\, \alpha(t) \leqq 1 + 0{,}04\sqrt{t}, \qquad t \leqq \bar{t} \qquad \text{(VI.3)}$$

$$\leqq 1 + c_1\sqrt{t}, \qquad t \leqq \bar{t}. \qquad \text{(VI.4)}$$

The behaviour of $\alpha(t)$ accounts for a flat pomeron for small t and incorporates the observed resonance on the f-trajectory for $t = m_f^2$, $t = m_h^2$, ... (cf. fig. 4).
In this simple model, the errors ε, ε^M, $\varepsilon(t)$, $\varepsilon^M(t)$ can be numerically estimated (using some aditional assumptions as discussed in appendix D). We will give the errors in the crossing

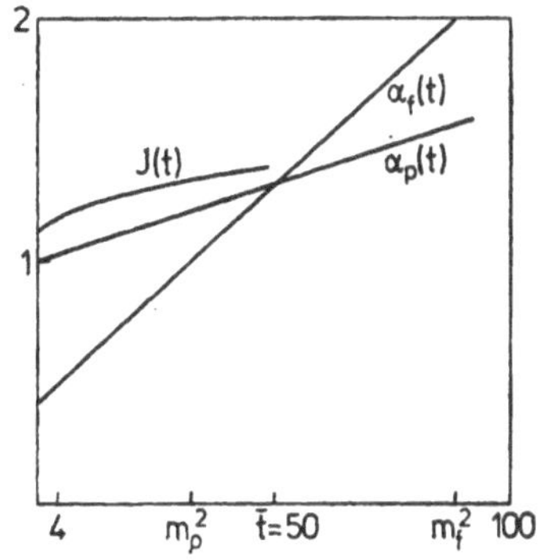

Fig. 4. Pomeron (α_P) and $f(\alpha_f)$-trajectories and the angular momentum $J(t)$ of the t-channel partial wave (VI.17)

sumrule (IV.17) with a weight function

$$\frac{c(s)}{s-4} = (b_0 + \lambda \ln s)^{1/2}\, s^{-\alpha_0-1} \tag{VI.5}$$

$$l = \frac{\sqrt{s}}{2}\,(b_0 + \lambda \ln s), \qquad J(t) = \alpha_0 + \lambda \sqrt{t}. \tag{VI.6}$$

In appendix D, the following bound on ε will be given (D.2, D.4)

$$|\varepsilon| \leqq 2\sqrt{2} \int\limits_{\bar t}^{\infty} dt\, \frac{\Gamma(J(t)+3/2)}{\Gamma(J(t)+1)}\, t^{-1/4} e^{-b_0\sqrt{t}} \left(\frac{t-4}{4}\right)^{-J(t)} \sqrt{\frac{t}{t-4}} = \varepsilon_1 . \tag{VI.7}$$

We choose values $\alpha_0 = 1{,}1$; $\lambda = 0{,}04$ $(= 0{,}057\ \mathrm{fm})$, $\bar t = 50$ $(0{,}9\ \mathrm{GeV}^2)$, such that $J(t)$ (cf. VI.6) stays near the pomeron trajectory (cf. fig. 4) for $4 \leqq t \leqq \bar t$. The value of $\bar t = 50 \simeq 4m_K^2$ is indicated, since the $\pi\pi$ interaction is effectively elastic below the $K\bar K$ threshold, and later on, elastic unitarity can be used for $t \in [4, \bar t]$.

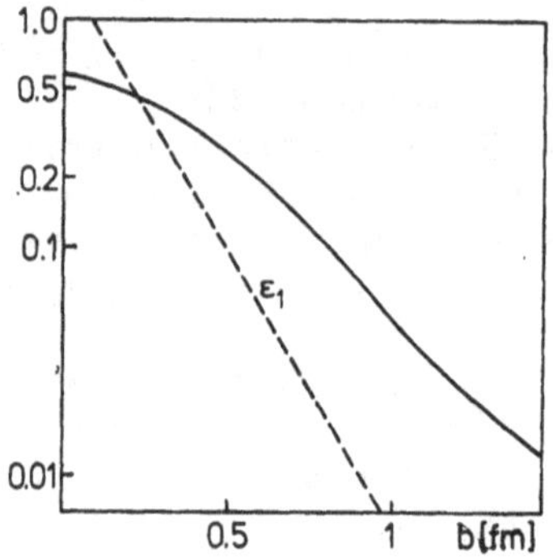

Fig. 5. The bound ε_1 (VI.17) (dashed line) for the error ε in (IV.17) as compared to the left hand side of the crossing-sumrule (IV.17) (solid line), calculated in the universal geometrical scaling model [*45*].

In fig. 5, the bound (VI.7) for ε is compared to a model calculation of the left hand side of the crossing sumrule, as a function of b_0. The $\pi\pi$ s-channel partial waves have been taken from the universal geometrical scaling model [*45*]. It is evident, that for $b_0 \geqq 1$ fm ε_1 becomes negligable. Therefore, s-channel partial waves with $b \geqq 1\ \mathrm{fm} + \lambda \ln s$ are determined by t-channel contributions in the elastic region $t \in [4, \bar t]$.
In the crossing sumrule, these t-channel contributions are approximated by the t-channel partial waves, and it is necessary to bound the approximation error ε^M in (IV.17). In appendix D it is shown (D.5—D.6), that

$$|\varepsilon^M| \leqq \int\limits_4^{\bar t} dt\, \gamma_1{}^M(t)\, \mathrm{Im}\, \bar f_{J(t)}(t), \tag{VI.8}$$

$$\gamma_1{}^M = 2\sqrt{2}\,\bigl(1024 \cdot (t-4)\bigr)^{(N-J)/2} \left(\frac{\Gamma(J+3/2)}{\Gamma(J+1)}\right)^2 \sqrt{J-N} \left(\int\limits_{\sigma(t)}^{\infty} ds\, s^{J+N+1} \varepsilon^M(s,t)^2\right)^{1/2} \tag{VI.9}$$

where $N = N(t) \geqq \mathrm{Re}\,\alpha(t)$ (cf. VI.2, IV.1). This bound has to be compared with the contribution of $\mathrm{Im}\, f_J(t)$ to the right hand side of the sumrule,

$$\int_4^{\bar{t}} dt\, a_0{}^M(t)\, \mathrm{Im}\, f_J(t)\,, \tag{VI.10}$$

where $a_0{}^M$ is defined in (IV.16).

$a_0{}^M$ and $\gamma_1{}^M$ have been calculated for $M = 0{,}1$, using the same values of parameters as in the discussion of ε. In figs. 6a–d they are compared for $t \in [4, \bar{t}]$ choosing $b_0 = 0.4$ ($= 0{,}585$ fm) and $b_0 = 1$ ($= 1{,}46$ fm). As can be seen, even for $M = 0$, ε^M will be less than 10% of the contribution (VI.10).

The important result of this quantitative error discussion is the following (if the Regge pole model and the other approximations, which are used in appendix D, are reasonable): The crossing sumrule connects s-channel partial waves, which describe the edge of the hadron ($b \geqq 1$ fm), to the t-channel partial wave $\mathrm{Im}\, \bar{f}_J(t)$ in the elastic domain $t \in [4, \bar{t}]$ up to an error which is less than 10%. $\mathrm{Im}\, \bar{f}_J(t)$ can be determined in terms of the central s-channel partial waves ($b \leqq 1$ fm) via dispersion relation (V.17) and elastic unitarity, with uncertainties $\varepsilon(t)$, $\varepsilon^M(t)$, which can be estimated in a similar manner as ε and ε^M above. This shows that the S-matrix principles analyticity and unitarity in the t-channel determine a rather large part of the b-distribution of s-channel partial waves. The results of this application of crossing sumrules and dispersion relation will be presented [*46*] elsewhere.

The foregoing considerations on the errors depend on the special model which had been chosen to represent the true amplitude in the domain $s \geqq 4$, $t \geqq 4$. In the following we

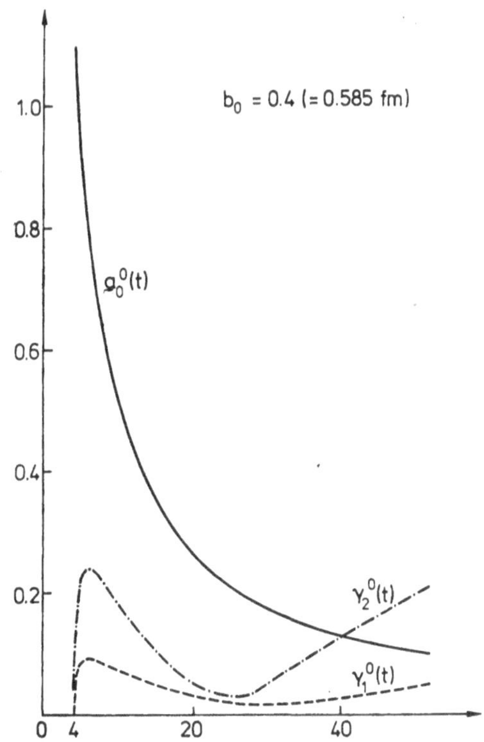

Figs. 6. Comparison of the first expansion coefficient $a_0{}^M(t)$ of (IV.17) (solid lines) with the error bounds $\gamma_1{}^M(t)$ (VI.9) (dashed lines) and $\gamma_2{}^M(t)$ (VI.14) (dashed-dotted lines)

a) $M = 0$, $b_0 = 0{,}585$ fm

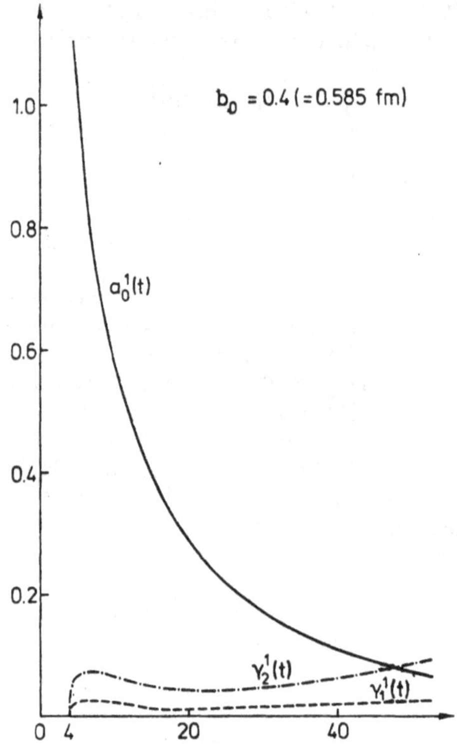

Fig. 6b.
b) $M = 1$, $b_0 = 0{,}585$ fm

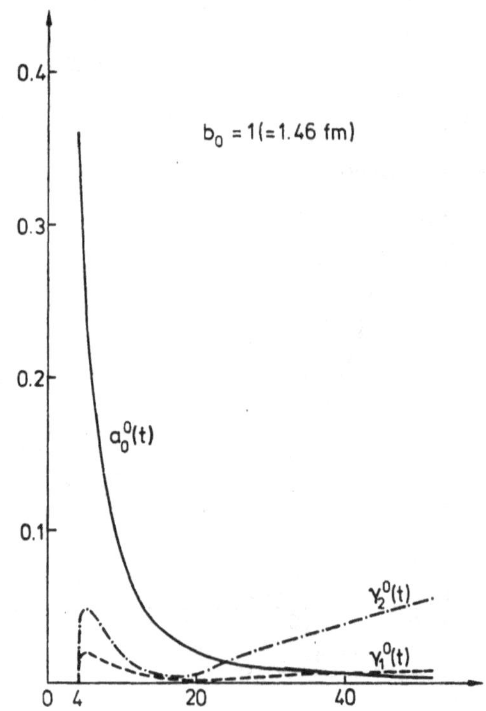

Fig. 6c.
c) $M = 0$, $b_0 = 1{,}46$ fm

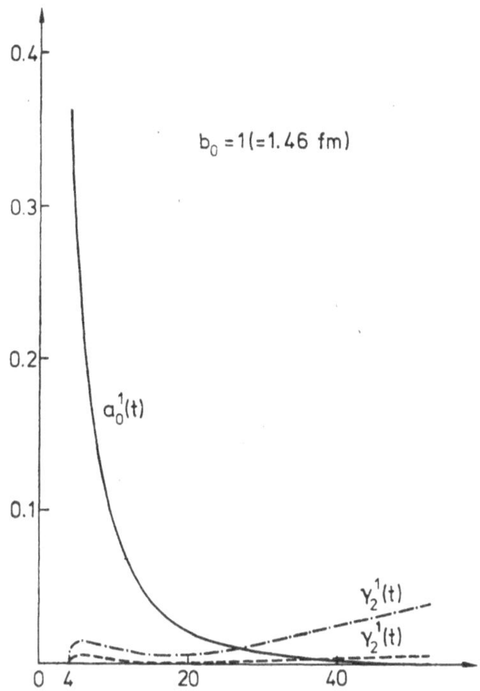

Fig. 6d.
d) $M = 1$, $b_0 = 1{,}46$ fm

will give different error bounds, which do not depend on a particular model for $A(s, t)$ but on considerations, which controll the oscillations of $\varrho(s, t)$.
The simplest and strongest property to require is the absence of oscillations

$$\varrho(s, t) \geqq 0 \quad \text{for} \quad s \geqq \sigma(t). \tag{VI.11}$$

This positivity guaranties automatically the positivity of all partial waves, which is a very desirable consequence. On the other hand, MARTIN [47] proved, that (VI.11) also implies a total cross-section which vanishes at least as $(\log s)^{-1}$.
Moreover, any Regge model with a complex trajectory has oscillations. Therefore, condition (VI.11) certainly is too strong to be valid for the true amplitude. Let us nevertheless for completeness give the error bounds, which can be derived from the positivity condition (VI.11). For the contribution (VI.1) to the Froissart-Gribov integral for s-channel partial waves one has

$$\varepsilon(s, l) = \frac{4}{\pi(s-4)} \int_{\bar{t}}^{\infty} dt Q_l \left(1 + \frac{2t}{s-4}\right) \varrho(s, t)$$

$$\leqq \frac{Q_l\left(1 + \frac{2\bar{t}}{s-4}\right)}{Q_{l_0}\left(1 + \frac{2\bar{t}}{s-4}\right)} \operatorname{Im} f_{l0}(s) = \varepsilon_2(s, l) \qquad l \geqq l_0 \tag{VI.12}$$

which follows from the positivity of $\varrho(s, t)$ and the monoton decrease of $Q_l(z)/Q_{l_0}(z)$ in z for $l \geqq l_0$, a property, which was proven in appendix B of ref. [34]. This bound is compared in fig. 7, for $s = 1{,}524 \cdot 10^5$ ($P_{\text{lab}} = 1480$ GeV), to the partial wave distri-

bution of the universal geometrical scaling model [45]. We choose $l_0 = \sqrt{s}/2\,\lambda\, \ln s$, $\lambda = 0{,}04$. In this case, too, the inelastic contributions $\varepsilon_2(s, l)$ from $t \geqq \bar{t}$ are small for $b \geqq 1$ fm.
The bound for ε^M under the positivity condition (VI.11) is derived in appendix D (cf. D. 7–9). It has the same form as (VI.8)

$$\varepsilon^M \leqq \int_4^{\bar{t}} dt\, \gamma_2{}^M(t)\, \mathrm{Im}\, \tilde{f}_J(t), \tag{VI.13}$$

$$\gamma_2{}^M(t) = \max_{s \geqq \sigma(t)} |\varepsilon^M(s, t)/\tilde{Q}_J(s, t)|. \tag{VI.14}$$

$\gamma_2{}^M(t)$ can be compared with $\gamma_1{}^M(t)$ and $a_0{}^M(t)$ in figs. 6.
Of course, one should look for a more realistic condition on $\varrho(s, t)$, which does not lead to contradictions with phenomenological results, but on the other hand still allows to give

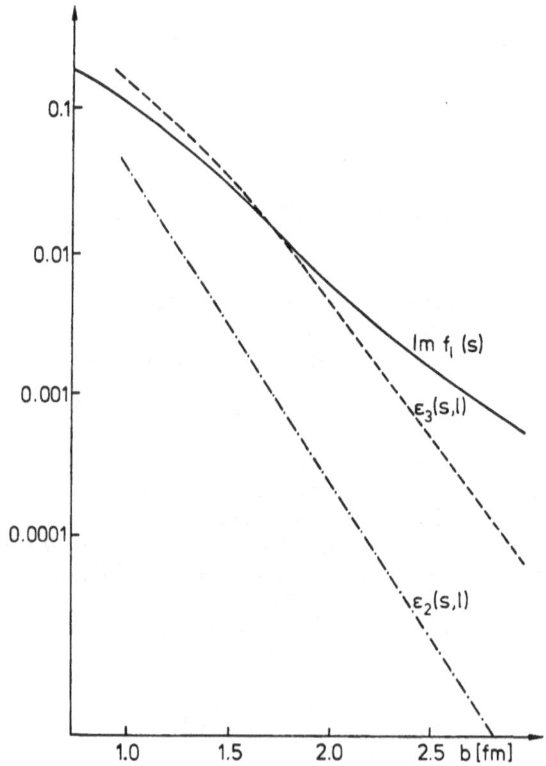

Fig. 7. Comparison of the s-channel partial waves in the model of ref. [45] with error bounds $\varepsilon_2(s, l)$ (VI.12) (dashed line) and $\varepsilon_3(s, l)$ (VI.23) (dash-dotted line) at $p_{lab} = 1480$ GeV

numerical error bounds. We will assume in the following that the amplitude can be split into two parts, a pomeron component and a f-component:

$$A(s, t) = A_p(s, t) + A_f(s, t). \tag{VI.15}$$

The f-component shall be a usual Regge pole contribution with intercept 0.5, which gives at large s vanishing contributions to s-channel partial waves. Therefore, $A_f(s, t)$ will be neglected in the following

$$A(s, t) = A_p(s, t). \tag{VI.16}$$

Furthermore, the Pomeron trajectory is assumed to be flat:

$$|A(s, t)| \leqq A s^{\mathrm{Re}\,\alpha_p(t)}, \qquad \mathrm{Re}\,\alpha_p(t) \leqq N(t) = 1 + \lambda_0 \sqrt{t}\,, \qquad t \geqq 4 \tag{VI.17}$$

with $\lambda_0 \simeq 0{,}04$ (cf. fig. 4).
A weaker condition than (VI.11) is the positivity of the integral over $\varrho(s, t)$:

$$\psi(s, t) = \frac{4}{\pi(s-4)} \int_{\sigma(s)}^{t} dt' Q_L \left(1 + \frac{2t'}{s-4}\right) \varrho(s, t) \geqq 0 \tag{VI.18}$$

for $s \geqq 4$, $t \geqq \sigma(s)$, where

$$L = \frac{\sqrt{s}}{2} \lambda_0 \ln s. \tag{VI.19}$$

There are several reasons to require such a property for the true amplitude ϱ:

i) It is known from the work of Mahoux and Martin [*48*] that $\varrho(s, t) \geqq 0$ in the domain (cf. fig. 1)

$$\sigma(s) \leqq t \leqq \frac{4\sqrt{s} + 8}{\sqrt{s} - 6} = \sigma_1(s). \tag{VI.20}$$

Therefore, $\psi(s, t)$ will certainly bepositive in an even larger domain than (VI.20). For $t \to \infty$, $\psi(s, t) \to \mathrm{Im}\, f_L(s)$ and $\psi(s, t)$ is positive in this limit, too. Condition (VI.18) requires that $\psi(s, t)$ stays positive also between these region. Thus (VI.18) allows oscillations of $\varrho(s, t)$, but controlls their magnitude. A qualitative picture of the possible behaviour of $\psi(s, t)$ is given in fig. 8.

ii) Condition (VI.18) guaranties the positivity of $\mathrm{Im}\, f_l(s)$ for $l \geqq L$, because

$$\mathrm{Im}\, f_l(s) = - \int_{\sigma(s)}^{\infty} dt \frac{d}{dt} \left[\frac{Q_l \left(1 + \frac{2t}{s-4}\right)}{Q_L \left(1 + \frac{2t}{s-4}\right)} \right] \psi(s, t) \tag{VI.21}$$

and $d/dt\, [Q_l/Q_L] \leqq 0$ for $l \geqq L$, according to appendix B of ref. [*34*].

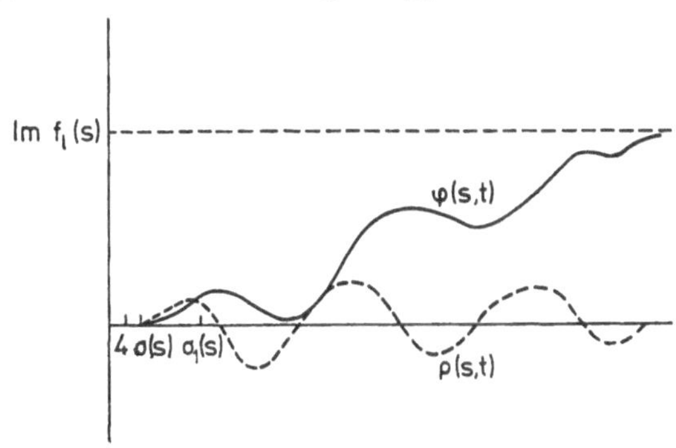

Fig. 8. An example of oscillating $\varrho(s, t)$ with positive $\psi(s, t)$ (VI.18)

iii) It is easy to show that (VI.18) is valid in Regge models, which fullfill the assumptions (III.24), (III.27) (which lead to the validity of the scaling law for $l \simeq \sqrt{s} \ln s$, cf. section III).

iv) It can be shown [49] that condition (VI.18) with L given by (VI.19) leads to the bound for the amplitude

$$|A| \leqq \text{const. } s^{1+\lambda_0\sqrt{t}}, \qquad t \geqq 4 \tag{VI.22}$$

with calculable const., which is consistent with the assumed Pomeron trajectory (VI.17).

The bound for $\varepsilon(s, l)$ which follows from the positivity of $\psi(s, t)$ is

$$\varepsilon(s, l) \leqq \left(\frac{Q_l(z)}{Q_L(z)}\right)' \left(\frac{Q_{l_1}(z)}{Q_L(z)}\right)'\Bigg|_{z=1+2t/s-4} \operatorname{Im} f_{l_1}(s) = \varepsilon_2(s, l). \tag{VI.23}$$

(VI.23) will be derived in appendix D (cf. D. 10—16). In fig. 7, this bound has been plotted as a function of b_0 for values of

$$l = \frac{\sqrt{s}}{2} b = \frac{\sqrt{s}}{2} (b_0 = \lambda \ln s)$$

$$l_1 = \frac{\sqrt{s}}{2} b_1 = \frac{\sqrt{s}}{2}(1/14 + \lambda \ln s) \tag{VI.24}$$

$$L = \frac{\sqrt{s}}{2} B = \frac{\sqrt{s}}{2} \lambda \ln s, \qquad \lambda = 0{,}04.$$

As can be seen in fig. 7, also the bound $\varepsilon_2(s, l)$ for $\varepsilon(s, l)$ drops much faster than the partial waves themselves, but its normalization is so large that it becomes small compared with $\operatorname{Im} f_l(s)$ only from $b = 2.5$ fm on.
The realistic condition on $A(s, t)$, which allows to estimate the contribution ε^M and $\varepsilon^M(t)$ to crossing sumrule (IV.17) and dispersion relation (V.17) is

$$\omega(s, t) = \frac{1}{\pi} \int\limits_{\sigma(t)}^{s} ds' \tilde{Q}_N(s', t)\, \varrho(s', t) \geqq 0 \tag{VI.25}$$

for

$$N(t) = 1 + \lambda_0 \sqrt{t} \geqq \operatorname{Re} \alpha_p(t), \qquad 4 \leqq t \leqq \bar{t}, \qquad s \geqq \sigma(t). \tag{VI.26}$$

The positivity of $\omega(s, t)$ can be motivated by the same reasonning as the positivity of $\varrho(s, t)$, (VI.16). (The validity of (VI.25) in Regge models was demonstrated in ref. [34]). The bound on ε^M of the crossing-sumrule can be calculated to give (cf. D. 17, 18)

$$|\varepsilon^M| \leqq \int\limits_4^{\bar{t}} dt\, \gamma_3{}^M(t) \operatorname{Im} \bar{f}_J(t), \tag{VI.27}$$

$$\gamma_3{}^M(t) = \max_{s \geqq \sigma(t)} \left|\frac{d}{ds}\, \varepsilon^M(s, t)/\tilde{Q}_N(s, t)\right| \left|\frac{d}{ds}\, \tilde{Q}_J(s, t)/\tilde{Q}_N(s, t)\right|^{-1}, \qquad J \geqq N. \tag{VI.28}$$

We have not included a comparison of $\gamma_3{}^M(t)$ with $a_0{}^M(t)$, since it turned out that the approximation of $\phi(s, t)$ with the methods of section IV and appendix B does not converge numerically in the norm $\gamma_3{}^M(t)$ for increasing M (whereas the norms $\gamma_1{}^M(t)$ and

$\gamma_2{}^M(t)$ decrease with M, as can be seen from the fig. 6). In order to use the positivity condition (VI.25), a modified approximation in terms of Legendre functions should be choosen which leads to a decreasing error $\gamma_3{}^M(t)$.

VII. Conclusions and Outlook

In the proceeding section we described a new method to exploit in the framework of S-matrix theory crossed channel properties of the elastic amplitude: analyticity and unitarity. The intention has been to find restrictions on the amplitude in the direct channel, expressed in terms of measurable quantities. This should allow to test and apply the consequences of crossed channel properties directly, without further dynamical input.

Such restrictions have been found for amplitudes having realistic analyticity and polynomial boundedness properties (in particular allowing rising Regge trajectories). It turned out that the peripheral s-channel partial waves $\mathrm{Im}\, f_l(s)$, i.e. partial waves with large angular momentum l, are restricted in terms of measurable quantities due to t-channel analyticity and elastic unitarity. The central s-channel partial waves, i.e. partial waves with small l, depend in addition on unknown inelastic t-channel contributions (Compare the construction of the full amplitude in refs. [*15*, *16*], where the inelastic part of the double spectral function is an almost arbitrary input).

The scaling law of section III. determines the asymptotic behaviour in l and s, using t-channel unitarity at threshold only. The normalization of the partial waves depends on the amplitude at the unphysical point $t = 4$, only the functional dependence on l and s is fixed. Restrictions on the absolute magnitude of the peripheral s-channel partial waves in terms of measurable quantities can be obtained by the use of the approximate crossing-sumrule of section IV, the dispersion relation of section V and of elastic unitarity in the t-channel. The main result of these sections was: Once the central partial waves with angular momenta $l \leqq \sqrt{s}/2\,(b_0 + \lambda \ln s)$ and t-channel partial waves with physical momenta $J = 2, 4, \ldots 2M$ in the elastic domain $t \in [4, \bar{t}]$ are given, one can calculate averages with respect to s of peripheral s-channel partial waves with angular momenta $l \geqq \sqrt{s}/2\,(b_0 + \lambda \ln s)$.

The proof of this statement involved an approximation of integrals over the double spectral functions in terms of t-channel partial waves. It is very convenient that the approximation scheme can directly be used to numerical determination of the averages of the peripheral partial waves.

So far the above metnioned problems, viz. to find restrictions on the elastic amplitude in terms of measurable quantities by exploiting crossed channel properties, has in principle been solved. However, these restrictions will be of phenomenological interest only, if not much input is needed, i.e. if the angular momentum $L = \sqrt{s}/2\,(b_0 + \lambda \ln s)$, which devides the amplitude in a central and a peripheral part is not too large, and if the number M of used t-channel partial waves is not too large. The number b_0 and M determined the uncertainty in crossing-sumrule and dispersion relation, which were due to the neglect of inelastic t-channel contributions and to the approximation of the double spectral function in terms of partial waves, resp. The rigorous statement was that both errors vanish in the limit $b_0 \to \infty$, $M \to \infty$. For finite b_0 and M, the bounds on the uncertainties depend in addition on the magnitude of the double spectral function. For a sensible application of the obtained results the uncertainties have to be controlled, which can be done up to now only by making additional assumption on the double spectral function. But it is important to keep in mind that these additional information concerns only the

errors in the crossing-sumrule and the dispersion relation and is not needed for the derivation of the relations themselves.
In section VIa numerical estimate of the uncertainties in the crossing-sumrule was performed, using a Regge pole model for the amplitude. In addition, more general positivity properties of the double spectral function are discussed, which also allow to control the errors. The errors turned out to be smaller than 10%, for $b_0 \simeq 1$ fm and $M = 0$ in the Regge pole model, for $b_0 \simeq 2$—3 fm and $M = 1$ under the positivity assumption. Thus these model calculations indicate that phenomenological interesting tests and applications of the consequences of crossed channel properties will be possible.
The next step will be an evaluation of a crossing sumrule and dispersion relation in $\pi - \pi$ scattering, where the central s-channel partial waves will be taken from some model (e.g. the universal geometrical scaling model [*45*]). The consistency of the model with the calculated peripheral partial waves can be checked and thereby the consistency of the model with t-channel properties. A related application is the derivation of upper bounds on the total cross-section, in the spirit of ref. [*51*], by constraining the central partial waves by s-channel unitarity and the peripheral ones by t-channel properties. Of course, here the error estimate is of special importance and the bounds are valid only for the particular model, which is used to determine the errors. These calculations are presently performed in collaboration with A. DE LA TORRE [*46*].
The most interesting applications of crossed channel properties, from a phenomenological point of view, will be the determination of the periphery of πN and NN scattering. Less rigorous use of t-channel properties (inserting a model πN amplitude in the Mandelstam equation for the NN double spectral function [*52*], showed a large disagreement with the data. Hopefully, it will be possible to reproduce the measured peripheral partial waves, taking into account analyticity and unitarity in the t-channel in a more systematic manner by using the methods which have been described in this review.

Appendix A

The asymptotic behaviour of Legendre functions.
For large values of l and $z \simeq 1$, the following representations for $P_l(z)$ and $Q_l(z)$ are convenient.

$$P_l\,(\cosh\alpha) = \frac{1}{\sqrt{2\pi}}\,\frac{1}{\sqrt{\sinh\alpha}}\,\frac{\Gamma(l+1/2)}{\Gamma(l+1)}\,e^{(l+1/2)\alpha}\big(1 + O(1/l\alpha)\big) \tag{A.1}$$

$$Q_l(\cosh\alpha) = \sqrt{\frac{\pi}{2}}\,\frac{1}{\sqrt{\sinh\alpha}}\,\frac{\Gamma(l+1)}{\Gamma(l+3/2)}\,e^{-(l+1/2)\alpha}\big(1 + O(1/l\alpha)\big). \tag{A.2}$$

These estimates are consequences of the expressions (3.2 (26) and 3.2 (44) or ref. [*40*]. In section III $l = x \cdot s^{3/2}$, $\cosh\alpha = 1 + 2t/(s-4)$. Therefore

$$\alpha = 2\sqrt{\frac{t}{s}}\,\big(1 + (2 - t/6)\,1/s\big) + O(s^{-2}). \tag{A.3}$$

These asymptotic behaviour of $\partial^{J+1}/\partial t^{J+1}\,P_l\big(1 + 2t/(s-4)\big)\big|_{t=4}$, of (III.3), is therefore

$$\frac{1}{\sqrt{\pi}}\,2^{-J-5/2}(x \cdot s)^{J-1/2}\,e^{4xs+(16/3)x}, \tag{A.4}$$

which leads to the scaling law (III.4). In ref. [27] only the first term in (A.3) had been treated, viz. $\alpha = 2\sqrt{t/s}$.
For $t = \bar{t}(s) = 4 + 32/\sqrt{s}$ α becomes

$$\alpha = 4/\sqrt{s} + 16/s - \frac{20}{3} s^{-3/2} + O(s^{-2}) \tag{A.5}$$

and the asymptotic form for Q_l

$$Q_l\left(1 + \frac{2\bar{t}(s)}{s-4}\right) \to \sqrt{\frac{\pi}{8}} \frac{1}{\sqrt{x}} s^{-1/2} e^{-4sx+16\sqrt{s}x+80/3x}. \tag{A.6}$$

Finally, for $t = \sigma(s) = 4 + 64/s - 16$

$$\alpha = 4/\sqrt{s} + \frac{112}{3} s^{-3/2} + O(s^{-2}), \tag{A.7}$$

which leads to

$$Q_l\left(1 + \frac{2t(s)}{s-4}\right) \to \sqrt{\frac{\pi}{8}} \frac{1}{\sqrt{x}} s^{-1/2} e^{-4sx-112/3x}, \tag{A.8}$$

$$\frac{Q_l\left(1 + \frac{2t}{s-4}\right)}{Q_l\left(1 + \frac{2\sigma(s)}{s-4}\right)} \to e^{96/3x - 2xs_1s_2 - 2x\tau}, \tag{A.9}$$

where

$$t = 4 + \frac{4s_1s_2}{s} + \frac{\tau}{s}. \tag{A.10}$$

For values of $l = \sqrt{s}/2\ (b_0 + \lambda \ln s)$ (A.2) yields

$$Q_l\left(1 + \frac{2t}{s-4}\right) = \sqrt{\frac{\pi}{2}} \sqrt[4]{t} \frac{e^{-b_0\sqrt{t}}}{\sqrt{t}\sqrt{b_0 + \ln s}} s^{-\lambda\sqrt{t}} \left(1 + O\left(\frac{1}{\ln s}\right)\right). \tag{A.11}$$

Appendix B

Approximation with Legendre functions

Here we give the justification of the approximation formula (IV.16). To begin with, we will sketch the proof for the completeness of the set (IV.15). Let us define, for $4 \leqq t \leqq \bar{t},\ 0 \leqq J,\ 0 \leqq x \leqq 1$

$$f_n(x) = (t-4)^{-J-2n-1} x^{-J-1} Q_{J+2n}\left(1 + \frac{2}{x(t-4)}\right). \tag{B.1}$$

According to eq. (A.3) of ref. [34], the $f_n(x)$ have the integral representations

$$f_{l+k}(x) = x^{2k}(t-4)^{-2l} \int_0^1 dz \big(z(1-z)\big)^{J+2k} \big(1 + z \cdot x(t-4)\big)^{-J-2k-1} W_1(z), \tag{B.2}$$

where the set $\{W_l(z), l = 0, \ldots\}$ is complete in the space $C\,[0, 1]$ of continuous function on the interval $[0, 1]$. Therefore it is possible to approximate by appropriate sums over l peaks $\delta(z - z_0)$ at z_0 under the integral

$$\sum_{l=1}^{N} a_{kl}(t-4)^{2l} f_{l+k}(x) \simeq x^{2k} \int_0^1 dz \big(z(1-z)\big)^{J+2k} \big(1 + zx(t-4)\big)^{-J-2k-1} \delta(z - z_0)$$

$$= \frac{\big(z_0(1-z_0)\big)^{J+2k}}{\big(1 + z_0 x(t-4)\big)^{J+1}} \left(\frac{x}{1 + z_0 x\,(t-4)}\right)^{2k}. \tag{B.3}$$

The set of powers $\{(x/1 + z_0x(t-4))^{2k}, k = 0, 1, \ldots\}$ is complete in $C[0, 1]$, this implies the completeness of the set $\{f_n(x), n = 0, 1, 2, \ldots\}$ out of which these powers can be built. For a rigorous derivation of this completeness property cf. the appendix of ref. [*34*]. What has been shown up to now is, loosely speaking the completeness of $\big(\tilde{Q}_J(s, t) = (t - 4/4)^{-J-1} Q_J(1 + 2s/(t-4))\big)$

$$\{\tilde{Q}_{J+2n}(s, t), \qquad n = 0, 1, \ldots\} \tag{B.4}$$

in the space of functions which vanish like s^{-J-1}, but what is needed is the completeness of the set (IV.15), i.e.

$$\{\tilde{Q}_J(s, t), \qquad \tilde{Q}_{2n}(s, t), \qquad n = 1, 2, \ldots\} \tag{B.5}$$

in the same space. But this is easy: Since

$$\{\tilde{Q}_{2n}(s, t), \qquad n = 1, 2, \ldots\} \tag{B.6}$$

is the set (B.4) with $J = 2$, and since

$$\tilde{Q}_{J+2n}(s, t) \to s^{-J-2n-1} \leqq s^{-3} \quad \text{for} \quad 0 \leqq J < 2, \qquad n = 1, 2, \ldots \tag{B.7}$$

every $\tilde{Q}_{J+2n}(s, t)$, for $0 \leqq J \leqq 2$, $n \geqq 1$, can be approximated by the set (B.6). Therefore the set (B.5) which is set (B.6) plus $Q_J(s, t)$ is as complete as the set (B.4). For the concret approximation problem (I.16) of section IV, it is convenient to introduce, for each fixed $t \in [4, \bar{t}]$, a Hilbertspace H_t of functions $f(s, t)$ with the scalar product

$$\big(f_1(., t), f_2(., t)\big)_t = \int_{\sigma(t)}^{\infty} ds s^{-J(t)-1+N(t)} f_1(s, t) \cdot s^{J(t)+1} \cdot f_2{}^*(s, t)\, s^{J(t)+1}, \tag{B.7}$$

where $J(t) = \alpha_0 + \lambda\sqrt{t}$, cf. (IV.13), and $N(t)$ is the power of the polynomial bound (IV.1) on the amplitude. Evidently the function $\phi(s, t)$, eq. (IV.10) and the approximation error $\varepsilon^M(s, t)$ in eq. (IV.16) are elements of H_t, and also the amplitude $\hat{A} = A(s, t) \cdot s^{-N(t)-J(t)-1}$, due to the bound (IV.1) belongs to H_t. Relation (IV.9) can be written in terms of the scalar product in H_t:

$$\int_{\sigma(\bar{t})}^{\infty} dsc(s) \operatorname{Im} f_l(s) = \frac{4}{\pi} \int_4^{\bar{t}} dt \big(\phi(\cdot, t), \hat{A}(\cdot, t)\big)_t + \varepsilon. \tag{B.8}$$

The set (B.5) is complete in H_t, for each $t \in [4, \bar{t}]$ (see appendix A or ref. [*34*]). Therefore, the theory of approximation in Hilbertspaces [*38*] can be applied, which allows to calculate the best coefficients $a_n{}^M(t)$, $n = 0, 1, \ldots M$ for the approximation (IV.16) from the set of $M + 1$ equations

$$\begin{aligned}(\phi(\cdot, t), \tilde{Q}_J(\cdot, t))_t &= a_0{}^M(t)\,(\tilde{Q}_J(\cdot, t), \tilde{Q}_J(\cdot, t))_t + \sum_{n=1}^{M} a_n{}^M(t)\,(\tilde{Q}_{2n}(\cdot, t), \tilde{Q}_J(\cdot, t))_t \\ (\phi(\cdot, t), \tilde{Q}_{2m}(\cdot, t))_t &= a_0{}^M(t)\,(\tilde{Q}_J(\cdot, t), \tilde{Q}_{2m}(\cdot, t))_t + \sum_{n=1}^{M} a_n{}^M(t)\,(Q_{2n}(\cdot, t), Q_{2m}(\cdot, t))_t .\end{aligned} \tag{B.9}$$

This choice of coefficients guaranties that the norm $\|\varepsilon^M(\cdot, t)\|_t$ of the error $\varepsilon^M(s, t)$ is minimal for given M and vanishes in the limit $M \to \infty$. Since the error ε^M in the sumrule (IV.19) can be estimated by

$$|\varepsilon^M| \leqq \frac{4}{\pi} \int_4^{\bar{t}} dt\, \|\varepsilon^M(\cdot, t)\|_t \cdot \|\hat{A}(\cdot, t)\|_t \leqq A \cdot \frac{4}{\pi} \int_4^{\bar{t}} dt\, \|\varepsilon^M(\cdot, t)\|_t \cdot \frac{\sigma(t)^{(N(t)-J(t))/2}}{\sqrt{J(t) - N(t)}} \tag{B.10}$$

the contribution ε^M vanishes in the limit $M \to \infty$.

For the derivation of the approximation (IV.20), we need the well known relation between Legendre and Bessel functions (see e.g. [39] for the derivation).

$$Q_l(\cosh\alpha) = K_0((l + 1/2)\,\alpha)\left(1 + O\left(\frac{1}{l}\right)\right). \tag{B.11}$$

For the Bessel function, the following integral representation holds [40]

$$K_0(z) = \frac{1}{2} \int_0^\infty e^{-z/2(y+1/y)}\, \frac{dy}{y}. \tag{B.12}$$

Therefore

$$K_0((l + 1/2)\,\alpha) = \frac{1}{2}\, e^{-(l+1/2)\alpha} \int_0^\infty \frac{dy}{y}\, e^{-(l+1/2)\alpha/2y}(y - 1)^2. \tag{B.13}$$

With $\cosh\alpha = 1 + 2s/(t - 4)$, $l = \sqrt{s}/2\, b = \sqrt{s}/2\,(b_0 + \lambda \ln s)$, (A.3) implies

$$(l + 1/2)\,\alpha = (b_0 + \lambda \ln s)\sqrt{t} + O\left(\frac{b}{s}\right). \tag{B.14}$$

Since, on the other hand, (compare relation (IV.12))

$$Q_J\left(1 + \frac{2s}{t - 4}\right) = \sqrt{\pi}\,\frac{\Gamma(J + 1)}{\Gamma(J + 3/2)} \left(\frac{t - 4}{4}\right)^{J+1} s^{-J-1}\left(1 + O\left(\frac{t - 4}{s}\right)\right), \tag{B.15}$$

the following crossing relation for Legendre functions holds.

$$\begin{aligned}\frac{s^{-\alpha_0}}{s - 4}\, Q_l\left(1 + \frac{2t}{s - 4}\right) &= \frac{1}{2\sqrt{\pi}}\, e^{-b_0\sqrt{t}} \\ &\int_0^\infty \frac{dy}{y} \exp\left[-\frac{b_0\sqrt{t}}{2y}(1 - y)^2\right] \frac{\Gamma(J + 3/2)}{\Gamma(J + 1)} \left(\frac{4}{t - 4}\right)^{-J-1} Q_J\left(1 + \frac{2s}{t - 4}\right)\Bigg|_{J=\alpha_0+\lambda\sqrt{t}(1+(1-y)^2/2y)} \\ &+ \bar{\varepsilon}(s, t)\end{aligned} \tag{B.16}$$

$$\bar{\varepsilon}(s, t) = \left(O\left(\frac{1}{\sqrt{s}\, b}\right) + O\left(\frac{b}{s}\right) + O\left(\frac{t-4}{s}\right)\right) \cdot \frac{s^{-\alpha_0}}{s-4} Q\left(1 + \frac{2t}{s-4}\right). \tag{B.17}$$

Appendix C

Properties of $h_L(s, t, t')$

The value of

$$h_L(s, t, t) = L\big(P_L(z)\, Q_{L-1}(z) - P_{L-1}(z)\, Q_L(z)\big), \qquad z = 1 + \frac{2t}{s-4} \tag{C.1}$$

shall be calculated.
The Wronskian of P_L and Q_L is (eq. 3.2(13) of ref. [*40*])

$$P_L(z)\, Q_L'(z) - Q_L(z)\, P_L'(z) = \frac{1}{1-z^2}. \tag{C.2}$$

Inserting into (C.2) the relation

$$P_L'(z) = -\frac{L}{1-z^2}\big(zP_L(z) - P_{L-1}(z)\big) \tag{C.3}$$

and a similar relation for $Q_L'(z)$, eq. (V.6) follows.
The behaviour of $h_L(s, t, t')$ at large s and $L = \left[\sqrt{s}/2\,(b_0 + \lambda \ln s)\right]$ can be calculated from

$$h_L(s, t, t') = \frac{L}{2}\big((P_L - P_{L-1})(Q_L + Q_{L-1}) - (P_L + P_{L-1})(Q_L - Q_{L-1})\big) \tag{C.4}$$

by using

$$P_L - P_{L-1} = \frac{2}{L}\sqrt{\frac{t}{s-4}\left(1 + \frac{t}{s-4}\right)}\, P_L^1 - \frac{2t}{s-4} P_L \tag{C.5}$$

and a similar formula for $Q_L - Q_{L-1}$. The expansions 3.2(28) and 3.2(45) of ref. [40] for Legendre functions now give

$$h_L(s, t, t') = \frac{1}{2}\left(\sqrt{\frac{\alpha}{\alpha'}} + \sqrt{\frac{\alpha'}{\alpha}}\right) e^{-L(\alpha'-\alpha)}$$
$$\times \left(1 + O\left(\frac{1}{L}\right) + O\left(\frac{1}{\alpha}\right) + O\left(\frac{1}{\alpha'}\right) + O\left(\frac{1}{L\alpha}\right) + O\left(\frac{1}{L\alpha'}\right)\right), \tag{C.6}$$

where

$$\cosh \alpha = 1 + \frac{2t}{s-4}, \qquad \cosh \alpha' = 1 + \frac{2t'}{s-4}.$$

The $t \leftrightarrow u$ exchanged term in (V.13) contains an extra power of $s^{-1/2}$. Inserting the estimate (A.3) for α and α' into (C.6), together with estimate (B.15) for $\tilde{Q}_J(s, t)$, yields expression (V.14) for $\phi(s, t, t')$.

Appendix D

Derivation of numerical error bounds

The residuum $A(t)$ in VI.2 is given in terms of the partial wave $\tilde{f}_J(t)$:

$$\tilde{f}_J(t) = A(t)\,\frac{1}{\pi}\int\limits_4^\infty ds\tilde{Q}_J(s, t)\; s^{\alpha(t)}$$

$$\cong A(t)\,\frac{1}{\sqrt{\pi}}\,\frac{\Gamma(J+1)}{\Gamma(J+3/2)}\,\frac{4^{-J+\alpha(t)}}{J-\alpha(t)} \tag{D.1}$$

only the leading s-power (B.15) of the Legendre function in the Froissart Gribov formula (D.1) has been used, since (VI.2) is a leading power model for the amplitude. For $l = \sqrt{s}/2$ $(b_0 + \lambda \ln s)$, $\lambda \geqq c_1$, ε can be calculated

$$\varepsilon = \int\limits_{\sigma(t)}^\infty ds\varepsilon(s, l)\cdot c(s) = \operatorname{Im}\frac{4}{\pi}\int\limits_{\bar{t}}^\infty dtA(t)\int\limits_4^\infty dsQ_l\left(1+\frac{2t}{s-4}\right)\frac{c(s)}{s-4}\,s^{\alpha(t)}$$

$$\cong \operatorname{Im}\,2\sqrt{2}\int\limits_{\bar{t}}^\infty dt\,\frac{\Gamma\big(J(t)+3/2\big)}{\Gamma\big(J(t)+1\big)}\,t^{-1/4}e^{-b_0\sqrt{t}}\tilde{f}_{J(t)}(t) \tag{D.2}$$

$$J(t) = \alpha_0 + \lambda\sqrt{t}\,. \tag{D.3}$$

The s-integration has been performed by approximating $c(s)/(s-4)\,Q_l\big(1+2t/(s-4)\big)$ by its asymptotic form (IV.12). Since in the simple Regge pole model (VI.2) $\tilde{f}_J(t)$ has J singularities only at complex values $J = \alpha(E)$ for $\bar{t} \geqq t \geqq 4$, the expression (D.2) for ε can be continued to real values $\lambda \leqq c_1$, i.e. $J(t) \leqq \operatorname{Re}\alpha(t)$, in particular to the value $\lambda = 0{,}04$ for which the bound will be calculated. Finally, the unitarity bound

$$|\tilde{f}_J(t)| \leqq \left(\frac{t-4}{4}\right)^{-J}\sqrt{\frac{t-4}{t}} \tag{D.4}$$

is assumed to hold for noninteger values of J.
This leads to the bound (IV.7) for ε.
The error ε^M is bounded by

$$|\varepsilon^M| \leqq \int\limits_4^{\bar{t}} dt\,\frac{1}{\pi}\left(\int\limits_{\sigma(t)}^\infty ds^{-J-1+N}\big(\varepsilon^M(s, t)\; s^{+F+1}\big)^2\cdot\int\limits_{\sigma(t)}^\infty dss^{-J-1+N}\big(s^{-N}\varrho(s, t)\big)^2\right)^{1/2}. \tag{D.5}$$

This is just Schwarz' inequality. Using the elastic unitarity equation (II.14) for $\varrho(s, t)$ and the estimates for the kernel $K(s, s_1, s_2, t)$ which are given in appendix A of ref. [*50*], one obtaines the bound for ϱ in terms of the model amplitude (VI.2)

$$|\varrho(s, t)| \leqq \frac{2}{\pi}\,S^N\left(\frac{t-4}{4}\right)^N\sqrt{\frac{t-4}{4}}\ln\frac{s(t-4)}{64}\,|A(t)|^2 \tag{D.6}$$

where $N = \text{Re}\,\alpha(t)$. Assuming furthermore, that the imaginary part of the trajectory is small in the domain $t \in [4, \bar{t}]$ (D.6) leads to expression (VI.9) for $\gamma_1{}^M(t)$.
The expression (VI.14) for $\gamma_2{}^M$, too, is derived using the unitarity equation (II.14) for $\varrho(s, t)$.

$$|\varepsilon^M| \leqq \int_4^{\bar{t}} dt \gamma_2{}^M(t) \frac{1}{\pi} \int_{\sigma(t)}^{\infty} ds \tilde{Q}_J(s, t)\, |\varrho(s, t)|, \tag{D.7}$$

$$\frac{1}{\pi} \int_{\sigma(t)}^{\infty} ds Q_J(s, t)\, |\varrho(s, t)| \leqq \sqrt{\frac{t-4}{4}} \left(\frac{t-4}{t}\right)^J \hat{f}_J(t)^2, \tag{D.8}$$

$$\hat{f}_J(t) = \frac{1}{\pi} \int_4^{\infty} ds \tilde{Q}_J(s, t)\, |A(s, t)| \leqq \frac{|J - \alpha(t)|}{J - N}\, |\tilde{f}_J(t)|. \tag{D.9}$$

The additional assumption of a small imaginary part of the trajectory leads in turn to the bound (VI.13).
The derivation of the bound for $\varepsilon(s, l)$ under the condition $\psi(s, t) \geqq 0$ (VI.18) makes more use of monotony properties of Legendre functions. From the definition (VI.1) for $\varepsilon(s, l)$ and (VI.18) for $\psi(s, t)$ a partial integration leads to

$$\varepsilon(s, l) = -\frac{Q_l(\bar{z})}{Q_L(\bar{z})}\, \psi(s, \bar{t}) - \int_{\bar{z}}^{\infty} dz \left(\frac{Q_l(z)}{Q_L(z)}\right) \psi(s, t) \quad l \geqq L, \tag{D.10}$$

where $\bar{z} = 1 + 2\bar{t}/s - 4$, $z = 1 + 2t/s - 4$.
Since $d/dz\,\big(Q_l(z)/Q_L(z)\big)$, for $l \geqq L$, is negative (cf. appendix B of ref. [*34*]) and $\psi(s, t)$ is positive, the following chain of inequalities hold:

$$\begin{aligned} \varepsilon(s, l) &\leqq Q_l(\bar{z})/Q_L(\bar{z}) \cdot \psi(s, \bar{t}) - M \int_{\bar{z}}^{\infty} dz\, \big(Q_{l1}(z)/Q_L(z)\big)' \cdot \psi(s, t) \\ &\leqq -M \int_{z_0}^{\infty} dz\, \big(Q_{l1}(z)/Q_L(z)\big)'\, \psi(s, t) = M\, \text{Im}\, f_{l_1}(s), \end{aligned} \tag{D.11}$$

$$l \geqq l_1 \geqq L, \qquad z_0 = 1 + \frac{s\sigma(s)}{s-4}$$

where

$$M = \max_{z \geqq \bar{z}} \big(Q_l(z)/Q_L(z)\big)' / \big(Q_{l_1}(z)/Q_L(z)\big)'. \tag{D.12}$$

Finally, the form (VI.23) of the bound for $\varepsilon(s, l)$ is established with the help of the following
Lemma: For $l \geqq l_1 \geqq L$, $z \geqq 1$, the function

$$h(z) = \left(\frac{Q_l(z)}{Q_L(z)}\right)' \left(\frac{Q_{l_1}(z)}{Q_L(z)}\right)' \tag{D.13}$$

is decreasing.
Proof: From the Legendre differential equation one obtaines

$$\begin{aligned} \frac{d}{dz} h(z) = \frac{1}{(z^2-1)} \frac{1}{\big(W_{l_1,L}(z)\big)} \big\{ &\big(l(l+1) - L(L+1)\big)\, Q_l(z)\, W_{l_1,L}(z) \\ &- \big(l_1(l_1+1) - L(L+1)\big)\, Q_{l_1}(z)\, W_{l,L}(z) \big\}\, Q_L(z) \end{aligned} \tag{D.14}$$

where (cf. appendix B of ref. 34)

$$W_{l,L}(z) = Q_l'(z)\, Q_L(z) - Q_L'(z)\, Q_l(z)$$
$$= -\frac{\big(l(l+1) - L(L+1)\big)}{z^2 - 1} \int\limits_z^\infty dz' Q_l(z')\, Q_L(z'). \qquad \text{(D.15)}$$

Therefore

$$\frac{d}{dz} h(z) = \frac{\big(l(l+1) - L(L+1)\big)\big(l_1(l_1+1) - L(L+1)\big)}{(z-1)^2 \big(W_{l_1,L}(z)\big)^2}$$
$$\times \int\limits_z^\infty dz' Q_L(z')\, Q_l(z')\, Q_L(z)\, Q_{l1}(z) \left(1 - \frac{Q_l(z)}{Q_{l_1}(z)} \Big/ \frac{Q_l(z')}{Q_{l_1}(z')}\right). \qquad \text{(D.16)}$$

Using once more that $Q_l(z)/Q_{l_1}(z)$ for $l \geqq l_1$ is monotonically decreasing, the integral in D.16 is seen to be negative, which establishes the lemma.
In a similar manner (VI.25) may be used to bound ε^M. By partial integration:

$$\varepsilon^M = -\int\limits_{\sigma(t)}^\infty ds \frac{d}{ds} \left[\frac{\varepsilon^M(s, t)}{Q_N(s, t)}\right] \omega(s, t). \qquad \text{(D.17)}$$

Since

$$\operatorname{Im} \tilde{f}_J(t) = -\int\limits_{\sigma(t)}^\infty ds \frac{d}{ds} \left[\frac{Q_J(s, t)}{Q_N(s, t)}\right] \omega(s, t), \qquad \text{(D.18)}$$

the condition $\omega(s, t) \geqq$ implies (VI.27).

References

[1] W. Heisenberg, Z. Phys. **120**, 513 (1943).
[2] D. Iagolnitzer, S-Matrix Theory: An Introduction (North Holland Publishing Company, Amsterdam 1976).
[3] Structural Analysis of Collision Amplitudes, ed. by R. Balian and D. Iagolnitzer, Les Houches June Institute 1975 (North Holland Publishing Company, Amsterdam, New York, Oxford 1976).
[4] M. Froissart, Phys. Rev. **123**, 1053 (1961).
[5] A. Martin, Scattering Theory: Unitarity, Analyticity and Crossing, Lecture Notes in Physics, Vol. 3 (Springer Verlag Berlin, Heidelberg, New York 1969).
[6] S. M. Roy, Phys. Rep. **5C**, **127** (1972).
[7] F. J. Yndurain, Rev. Mod. Phys. **44**, 645 (1972).
[8] A. Martin, preprint TH-2229-CERN (1976).
[9] S. M. Roy, Phys. Letters **36B**, 353 (1971).
[10] B. R. Martin, D. Morgan, G. Shaw, Pion-Pion Interaction in Particle Physics (Academic Press, London 1976).
[11] S. Ciulli, C. Pomponin, I. Sabba-Stefanescu, Phys. Rep. **17C**, 133 (1975).
[12] H. Lehmann, Nuovo Cimento **10**, 579 (1958); A. Martin, Nuovo Cimento **42**, 930 (1966).
[13] S. Mandelstam, Phys. Rev. **112**, 1344 (1958).
[14] For a review and references to the original literature, cf. Chs. III, VI, of P.D.B. Collins and E. J. Squires, Regge Poles in Particle Physics, Springer Tracts in Modern Physics **45** (1968).
[15] D. Atkinson, Springer Tracts in Modern Physics **57**, 1 (1971), and p. 415–426 cf. ref. [3].

[16] J. KUPSCH, On the Existence of Crossing Symmetric and Unitary Scattering Amplitudes with Regge Poles, Technical Report, Univ. Kaiserslautern (May 1977), and references therein.
[17] H. D. I. ABARBARNEL, J. B. BRONZAN, R. L. SUGAR, A. R. WHITE Phys. Rep. **21C**, 120 (1975).
[18] G. SOMMER, Fortschr. Phys. **18**, 577 (1970).
[19] G. F. CHEW, The Analytic S-Matrix (W. A. BENJAMIN, New York 1966).
[20] J. BROS, H. EPSTEIN, V. GLASER, Nuovo Cimento **31**, 1265 (1964).
[21] H. EPSTEIN, V. GLASER, A. MARTIN, Comm. Math. Phys. **13**, 257 (1969).
[22] Y. S. JIN, A. MARTIN, Phys. Rev. **135B**, 1375 (1964).
[23] G. AUBERSON, T. KINOSHITA, A. MARTIN, Phys. Rev. **D3**, 3185 (1971).
[24] V. DE ALFARO, T. REGGE, Potential Scattering (North Holland Publishing Company, Amsterdam 1965).
[25] D. AMATI, S. FUBINI, A. STANGHELLINI, Nuovo Cimento **26**, 896 (1962).
[26] MANDELSTAM, S., Phys. Rep. **13C**, 259 (1974).
[27] O. HAAN, K. H. MÜTTER, Phys. Letters **52B**, 472 (1974), **53B**, 1 (1974).
[28] A. K. COMMON, Phys. Letters **55B**, 318 (1975).
[29] F. J. YNDURAIN, Nuclear Phys. **B 92**, 51 (1975).
[30] P. GRASSBERGER, A. MARTIN, unpublished.
[31] A. K. COMMON, Nuclear Phys. **B 106**, 176 (1976).
[32] O. HAAN, K. H. MÜTTER, Nuclear Phys. **B 94**, 365 (1975).
[33] J. DIAS DE DEUS, P. KROLL, Phys. Letters **60B**, 375 (1976).
[34] O. HAAN, K. H. MÜTTER, Nuclear Phys. **B 131**, 232 (1977).
[35] B. SCHREMPP, F. SCHREMPP, Nuclear Phys. **B 77**, 453 (1974).
[36] K. H. MÜTTER, unpublished.
[37] K. H. MÜTTER, unpublished.
[38] N. I. ACHIEZER, Theory of Approximation (Fredrik Ungar Publishing Company, New York 1956).
[39] G. SZEGÖ, Orthogonal Polynomalis, American Math. Society, Colloquium Publications Vol. 23 (Providence, Rhode Island, 1939).
[40] Bateman Manuscript Project: Higher Transcendental Functions, Vol. 1, 2, 3, ed. by A. Erdélyi (Mc Graw-Hill Book Company, New York 1953).
[41] O. HAAN, Preprint WU-B 77–11, Univ. Wuppertal (1977).
[42] A. MARTIN, Nuovo Cimento **61A**, 56 (1969).
[43] Eqs. 3.8 (2), (21) of ref. [40].
[44] Cf. ch. V of ref. [14].
[45] J. DIAS DE DEUS, P. KROLL, Nuovo Cimento **37A**, 67 (1977).
[46] O. HAAN, A. DE LA TORRE, in preparation.
[47] Cf. ref. [42].
[48] G. MAHOUX, A. MARTIN, Nuovo Cimento **33**, 883 (1964).
[49] O. HAAN, unpublished.
[50] J. KUPSCH, Comm. Math. Phys. **19**, 65 (1970).
[51] F. J. YNDURAIN, Phys. Letters **31B**, 368, 620 (1970); A. K. COMMON, Nuovo Cimento **69A**, 115 (1970).
[52] J. W. ALCOCK, W. N. COTTINGHAM, C. MICHAEL, Nuclear Phys. **B 67**, 445 (1973).

INHALT

Die „FORTSCHRITTE DER PHYSIK“ dienen der Publikation zusammenfassender Berichte über aktuelle Teilgebiete der Physik. Neben Originalberichten in deutscher oder englischer Sprache werden deutsche Übersetzungen wichtiger fremdsprachiger Berichte abgedruckt.

Bei der Schriftleitung eingegangene Arbeiten:

U. NIEDERER: Massless Fields as Unitary Representations of the Poincaré Group

L. HECK: The Algebraic Structure of the Thirring Model

B. GEYER, D. ROBASCHIK, E. WIECZOREK: Theory of Deep Inelastic Lepton-Hadron Scattering

M. CARMELI, M. KAYE: Einstein-Maxwell Equations: Gauge Formulation and Solutions for Radiating Bodies

H. BAIER: Exchange Currents, Isobaric Excitations and the Deuteron

Manuskripte werden nach Vereinbarung mit dem Autor von der Schriftleitung entgegengenommen.

FSC
www.fsc.org
MIX
Papier | Fördert
gute Waldnutzung
FSC® C083411

Zeitfracht Medien GmbH
Ferdinand-Jühlke-Straße 7
99095 Erfurt, Deutschland
produktsicherheit@kolibri360.de